THE FUTURE IN THE PRESENT

HOW AI WILL IMPACT YOUR LIFE

TONY TAN

PARTRIDGE

Copyright © 2022 by Tony Tan.

ISBN: Hardcover   978-1-5437-6895-4
       Softcover   978-1-5437-6893-0
       eBook   978-1-5437-6894-7

All rights reserved. No part of this book may be used or reproduced by any means, graphic, electronic, or mechanical, including photocopying, recording, taping or by any information storage retrieval system without the written permission of the author except in the case of brief quotations embodied in critical articles and reviews.

Because of the dynamic nature of the Internet, any web addresses or links contained in this book may have changed since publication and may no longer be valid. The views expressed in this work are solely those of the author and do not necessarily reflect the views of the publisher, and the publisher hereby disclaims any responsibility for them.

Print information available on the last page.

To order additional copies of this book, contact
Toll Free +65 3165 7531 (Singapore)
Toll Free +60 3 3099 4412 (Malaysia)
orders.singapore@partridgepublishing.com

www.partridgepublishing.com/singapore

This book is dedicated to everyone who is looking to harness AI and digital innovations to optimise every aspect of their lives.

To work together as a community of global digital nomads to create a better, healthier, sustainable, and happier planet through mindful adoption of cutting-edge technologies.

OPENING QUOTE

To avoid bad outcomes, for every dollar and every minute we invest in improving artificial intelligence, it would be wise to invest a dollar and minute in advancing human consciousness.

Yuval Noah Harari

TESTIMONIALS

Our potential lies in the future we create. Tony Tan is a visionary entrepreneur who has a clear idea of what the future can be, and what it will be, if we simply slide into it, rather than creating it by choice. Tony presents possibilities and strategies that put the power back into our hands. This book is a must-read for anyone who wants to optimise their life and their future.

Vikas Malkani
The Wisdom Coach
Founder of SoulCentre®, SoulKids® and W.O.W. Coaching™

Tony is an inspiring business leader who is always at the cutting edge of technology and innovation, who inspires his teams to deliver excellence and always overdelivers in terms of value and professionalism.

A lifetime learner, Tony is a 360 entrepreneur with both external business skills and an inner depth of wisdom. *The Future in the Present* will be a game changer for many, and is much needed in this fast-paced, evolving world. I highly recommend it to all readers.

Sally Forrest
BSc, MRPharmS, MBA
CEO, SoulCentre Pte Ltd and TEDEx Speaker

As a futurist and technology leader, Tony Tan is the primary translator in the field of artificial intelligence for enterprises and individuals. His research and understanding of where the future is going in technology and AI is unmatched. He is a thought leader in digital innovations, and is unrivalled in his ability to see the future in the present.

Sam Cawthorn
11-time bestselling author
Australian of the Year
Founder and Director, Speakers
Institute Group of Companies

This century's must read! I love this point from the book: "To master technology you must first master the mind." Understanding that AI is delivering a new future today and how it is optimising our lives, giving us back freedom and saving us time is a great thing. Thank you for bringing this book to life—I am securing my seatbelt as the future enters the present!

Catherine Molloy
Global Keynote Speaker
Bestselling author, *The Million Dollar Handshake* and *The Conscious Leader*

Tony Tan's book offers readers an insight into one of today's paradigm shifting topics. Always able to look ahead, Tony helps navigate the reader into the future from where they stand today, helping both businesses and individuals alike with his thought-provoking approach. I'm honoured to have been inspired and advised by his work.

Dr Vikas Kumar Singh
International marketing strategist and speaker

Tony is my go-to advisor on AI and technology for business. His creativity and disruptive ideas are provocative and inspiring.

Michelle Lim
Social Entrepreneur, Edupreneur and TEDEx Speaker
CEO, MAD School and Make the Change (B Corp)

Tony is a leading futurist who not only speaks widely on digital revolutions, but runs his own award-winning AI organisations. I am confident that *The Future in the Present* is going to be a key book to unlock the potential of AI.

Kenneth Lai
Area Vice President, Splunk

Tony Tan always explores, ventures and reinvents himself to stay ahead—so he can bring a piece of the future back to the present.

Stanley Toh
Head, Enterprise End-users Services and Experience
Broadcom Limited

Congratulations on the upcoming publication of your book on AI. Your vision is inspiring, and few have the courage to pin down the future of transcendence. After all the hard work you have put in, I am looking forward to seeing your book on the bestsellers' list!

Franco Chiam
Head of Digital Workplace for Global IT
ASM International

Tony didn't need to write a book, but I am so glad he did. The views he shares in this book as a tech futurist on digital innovations and AI will definitely be insightful for many of us.

Boey Chern-Yue
Senior Vice President, SailPoint

AI drives the future of technology. It is powerful, but it does not replace love, care and human goodness. It will be a quantum leap for AI to reach such a level in an unforeseeable future. This thought-provoking book by Tony discusses how technology will evolve through AI and plant the seeds to explore our humanity in this tech revolution. Let's collectively play our part and steer it in the right direction!

Christopher Lek
Director, Cyber Security
Centre for IT Services, National
Technological University, Singapore

Tony is a futurist who not only understands the capabilities and limitations of AI and digital disruptions, but also what the future holds for the coming Metaverse. *The Future in the Present* is a must-read for anyone who wants to optimise their career and life with technology augmentation. Highly recommended!

Benjamin Pheong
Security Services Manager, IBM GCG

CONTENTS

Foreword: AI and Us xvii
Introduction: Welcome to the Future xxi

Chapter 1   On the Shoulders of Giants 1
Chapter 2   You Can't Spell Gaia Without 'AI' 16
Chapter 3   Gold Solutions for a Silver Population 28
Chapter 4   On the Trail of a Pandemic 39
Chapter 5   AI Comes Home: The Internet of Things ... 49
Chapter 6   The Weapons of the Future 62
Chapter 7   The Urgency of Cybersecurity 74
Chapter 8   Smart Singapore 85
Chapter 9   The Foundation of Hope 98

Conclusion: The Human Mind, Unleashed 125
Afterword: AIs and Conscious Thought 133
Acknowledgements 141
About the Author 143
About the Co-Author 145

FOREWORD

AI and Us

Being someone who lives and breathes technology, I've always been a fan of bringing it to the masses to improve our everyday lives. When Tony told me of his book to clarify AI, digital innovations, and their uses for the betterment of humanity, I was intrigued.

Everyone has different definitions and perspectives when it comes to AI applications. One camp, including the late physicist Stephen Hawking and pioneering entrepreneur Elon Musk, warns that someday soon, AI will be used to cause great harm to humanity.

Another, more optimistic group points to it as a force multiplier, playing a dominant role in augmenting humanity and vastly improving our innovation and productivity. The latter camp includes Lee Kai-Fu, former president of Google China, whose career of over forty years has been spent in the field; and Ray Kurzweil, the author, inventor, and futurist who has written much on how AI will empower, rather than replace, us.

That said, the weaponisation of AI by cybercriminals and nation-states is real and concerning. AI-powered codes, robots, and drones will play a bigger part in the future of crime and warfare. AI is just like any other technology—it is neither good nor bad. Humanity must deploy it mindfully and wisely, in the same way we do every advance from fire to nuclear energy. When deployed and regulated rightly, AI's potential can be employed to provide better cybersecurity, reduce collateral damage in conflicts, and better track down criminals to the benefit of society. It can also help resolve larger issues like climate change, finding cures to diseases, and predicting future pandemics.

Of course, TV programmes and movies often depict AI as a superintelligence that tries to wipe out humanity. Real life is not so dramatic, though it is indeed still a tool for those seeking to spread fake news, manipulate elections, or worse.

The point is that AI is here to stay; it's already creating massive benefits by optimising human potential and creating value, but the issue is what ends this ability will be turned to. AI is an omni-use technology and is already infused into the world of digital commerce, logistics, finance, retail, and transportation.

According to a report by McKinsey, AI is to deliver additional global economic activity of around $13 trillion by 2030, or about 16 per cent higher cumulative GDP compared with today. This amounts to 1.2 per cent additional GDP growth per year.[1] With so much at stake, it's clear that how we use these

---

1   Jacques Bughin, Jeongmin Seong, James Manyika, Michael Chui, and Raoul Joshi, 'Notes from the AI frontier: Modeling the impact of AI on the world economy', *McKinsey*, 4 September 2018, at https://www.mckinsey.com/featured-insights/artificial-intelligence/notes-from-the-ai-frontier-modeling-the-impact-of-ai-on-the-world-economy.

technologies, now and in the future, will determine the course of humanity's progression.

Tony reminds us: 'To master technology, you must first master your mind.' Having the right mental models to deploy AI mindfully and consciously will be critical in steering towards a positive outcome.

This is where Tony and Vikas Malkani go beyond other books on AI. Those focus on or predict AI use cases, but Tony and Vikas go behind the scenes and look at how the right mentality towards AI helps us to maximise its benefits.

I've experimented with AI and neural computing since the early 1990s and seen various successes and failures. What Tony has done is bring together the various aspects of AI adoption and the thoughts of industry leaders with a more holistic view of the way forward, which we'll need as we develop Singapore into a smart nation.

I hope that you will gain a better appreciation of his passion for AI after reading this book and, should you enter this exciting field, that you look on the technology in your hands with a new perspective . . . and the respect you'll need for both it and your fellow man.

Ian Loe
CTO, NTUC Enterprise
Adjunct Senior Fellow, Singapore University
for Technology and Design

INTRODUCTION
Welcome to the Future

Before the Romans came, the Israelite in Bethlehem or the Syrian in Palmyra lived out his days dissevered from, and in fact in ignorance of, the wider world. His universe ended at the town gate or the communal well. Could he trade? Study? Venture abroad? How, when he could know no more of the world than he could see from his doorstep or make plans for the morrow no farther than the distance he could tramp today?

Rome brought the mail, and the mail brought the world.

Steven Pressfield, *A Man at Arms*

Imagine for a moment sharing the same birthday as powered flight—17 December 1903. You don't know it yet, but your first cries happen at the same time as Orville and Wilbur Wright made the first successful heavier-than-air powered flight at Kitty Hawk, North Carolina.

Against all the odds, you make it out of early childhood spared from the infections that claim or debilitate large numbers of your peers and their mothers.

At ten years of age, if you were well-educated—a privilege granted only to a very few, usually nobility and the children of the well off—you would have heard of powerful states like the British Empire, that world-spanning territory on which the sun never set, the declining but still formidable Ottomans, Austro-Hungary, and more.

As you're living through your teens, an assassin ignites this powder keg of shifting powers, feuds, and alliances by shooting the archduke of Austro-Hungary dead on a Sarajevo street. For some reason that scholars spend decades unravelling, a devastating five-year war breaks out and consumes the European continent and its colonies, claiming millions of lives and tearing those great powers down in the blink of an eye.

As if the tremendous suffering and death in the trenches of Europe in what will be known as the First World War wasn't bad enough, a new strain of influenza sweeps across the world, leaving millions more dead. A man might survive the horrors of war, only to return home and enjoy peacetime for a short while . . . before being felled by this modern plague.

Before you're thirty, the Great Depression has destroyed large amounts of wealth, perhaps leaving you nearly penniless. Before you're forty, the worst war in human history has broken out, claiming hundreds of millions more lives. It isn't unusual for a family to lose multiple family members in a short span of time. Yet technology marches on, and your own life has seen the airplane go from a mere curiosity to a crucial tool of travel and warfare. What will be known as the Second World War triggers even more technological advancement, including the splitting of the atom, nuclear weapons, and the destruction of entire cities with a single blow. You hear it argued that for the first time, mankind has the power to end its own existence.

But the century has more in store. By the time you're in your fifties, people are flying into space. Before you're out of your sixties, the Apollo programme has put men on the moon. While entire millennia have gone by with very little change, it has now taken less than a single human lifetime to go from Kitty Hawk to the moon.

But that is merely the tip of a very large iceberg. Down on earth, life expectancy has been increasing as food production skyrockets, computers shrink in size and grow in power, and new advances in medical science, civil engineering, sanitation, and more ensure more people grow up healthier and live longer, more productive lives.

If you're blessed with a normal lifespan in the developed world, you'll live to see the rise of the Internet, and computers going from a luxury only the largest organisations could afford to cheap, commonplace devices in practically every home. The world hasn't just changed over the course of your life—it's flattened, as the ordinary gain powers once available only to the elite. It's shrunk, as geographical distance does less and less to slow the speed of communication; your life has seen the telegraph give way to phone conversations, email, and instant messaging. *The* world may be smaller, but *your* world has only grown. More for you than ever, it's a grand, wondrous place to explore.

Your children will see the same thing happening in their time, and what they saw invented and built over the years, your grandchildren will instinctively build into their lives. What you saw happen with computers, they'll see happening with the Internet and mobile phones, as the lines between devices fizzle, blur, and disappear altogether. They'll also enjoy vaccination and immunity from many a disease that plagued your time, like measles, the dreaded polio, or new outbreaks of respiratory illness. For all the media buzz around 2003's SARS pandemic or the present COVID-19 one, there is simply no comparison between them and the 50 million-strong death toll of the Spanish flu a century ago.

For the vast majority of human history, such rapid increases in prosperity were unheard of, and in many societies where it has passed by, it is still the case today. A farmer in medieval England had little reason to believe life would be any different for his grandchildren, if plague, warfare, or an accident did not take them first. (All three are still with us, but today they are far more survivable.) Today's young people have far more choices and information to make them than our ancestors ever did, and the things they'll be empowered to do can only be guessed at.

Those are the guesses we'll be making in this book as we try to make some sense of the technological possibilities that await us in an increasingly automated world—one where artificial intelligences (AIs) make more and more decisions for us. Will they bring us closer to the wondrous future imagined in, say, *Star Trek*, with its holodecks and material prosperity for all? Or will we fall into darkness and find ever more ways to destroy one another, like in *Black Mirror* or *Dr. Strangelove*?

As the nuclear age has shown us, there are deadly consequences if we get this wrong. The more capable and useful a new tool is for good, the worse and more destructive it will be when turned to the wrong ends. Ever since early man began setting fires and building weapons, we've known this principle. The same fires that kept our ancestors warm (and enabled them to see at night and cook their food) also destroyed entire forests. The same dynamite that lets us mine the earth and blast new paths through solid rock can level entire cities, and the same computers that give us such convenience can allow hackers and identity thieves to take over our lives.

Science fiction poses questions, and history can go some way towards answering them. In this book, we'll be exploring several reasons for this, and preparing you, the reader, for a future that's both more uncertain than ever but far more exciting to explore and live in.

It's on this journey that we'll discover causes for both optimism and pessimism (and learn why we'll always need both), the need for a high view of humanity that science and engineering alone cannot provide, and how we can choose the good and avoid the bad so that life gets as good as it can, for the most people possible.

Five Industrial Revolutions

'Imagine you're an alien assigned with keeping tabs on *Homo sapiens* over the last 250,000 years. Every 10,000 years you check in.' So begins US commentator Jonah Goldberg's book *Suicide of the West*, on the economic, industrial, and social revolutions he terms an incredible miracle.

The first twenty-three times, you'd record the same thing in your notebook: 'Semi-hairless, upright, bands of nomadic apes foraging and fighting for food. No change.' But everything changes on the twenty-fourth:

> Basic agriculture and animal domestication have been discovered by many of the scattered human populations. Some are using metal for weapons and tools. Clay pottery has advanced considerably. Rudimentary mud and grass shelters dot some landscapes (introducing a new concept in human history: the *home*). But there are no roads, no stone buildings worthy of the label. Still, a pretty impressive advance in such a short period of time, a mere 10,000 years.
>
> Eagerly returning 10,000 years later, our alien visitor's ship would doubtless get spotted by NORAD.[2]

Goldberg's point is that nearly *all* human progress—from nomadism to the homes we take for granted—has happened in what amounts to an eye-blink in the sweep of history. Human

---

2   Jonah Goldberg, *Suicide of the West: How the Rebirth of Tribalism, Populism, Nationalism, and Identity Politics Is Destroying American Democracy* (New York, New York: Crown Publishing Group, 2018), 'Introduction'. NORAD is the North American Aerospace Defense Command, the combined US and Canadian organisation that provides aerospace surveillance and protection for northern America.

prosperity swept across the world, to the extent that we today live better than even kings did a century ago. In medieval times, a king might host a banquet with hundreds of courses for his guests, all prepared by an army of cooks, servants, and other staff in his employ. Today, any one of us enjoys that privilege—we can simply walk into a restaurant, food centre, or cafe of our choice, and eat what we want. That's right, in any of Singapore's hawker markets, one can literally eat like a king on very little money.

Goldberg points to the invention of money as a key step in world progress because 'it lowers the barriers to beneficial human interaction'.

> It reduces the natural tendency to acquire things from strangers through violence by offering the opportunity for commerce. A grocer may be bigoted toward Catholics, Jews, blacks, whites, gays, or some other group. But his self-interest encourages him to overlook these things. Likewise, the customer may not like the grocer, but the customer's self-interest encourages her to put such feelings aside if she wants to buy dinner. In a free market, money corrodes caste and class and lubricates social interaction.

As people formed villages, towns, and cities with their own banks, with money circulating in their own economies, there came competition for resources—and as the saying goes, necessity is the mother of invention. That led to several industrial revolutions, each one made possible by the one before it:

1. *The spread of knowledge.* Printing came to Europe when Johannes Gutenberg invented the first printing presses, allowing books like the Bible and great classical works to be quickly copied and spread around the continent. Reading and writing exploded in popularity, and the knowledge of past generations could be recorded and learnt by not only the wealthy but also ordinary people. It's easy to take literacy for granted, but up until a few generations ago, only the elite could teach their children how to read and write.

   Suddenly, attempts by the authorities to stifle knowledge came to naught. Where governments and sometimes the church tried to prevent translations of the Bible and other valuable books from reaching the masses, demand for them kept presses turning and smugglers in business. No matter how many writers, translators, or publishers were arrested, vernacular books appeared all over Europe, and the continent was never the same again.

2. *The spread of steam power.* As machines grew more advanced and engineers built on existing designs, steam power made the back-breaking work of farming and transporting goods across vast distances things of the past—through steam engines, steamships, and steam locomotives. Agrarian societies became urbanised, and new methods of manufacturing created new changes to daily life. Mass production and assembly lines became the norm, allowing more and more to be made and put into the hands of people.

During this time, the electrical telegraph revolutionised communications around the world. By the late nineteenth century, communication lines linked the world-spanning British Empire, allowing news to be carried all over the globe.

3. *The spread of convenience.* A third industrial revolution came just before World War I, which saw numerous new inventions—such as the telephone, the electric light bulb (which opened up new possibilities of working at night), and the automobile.

All those things replaced difficult, manpower-intensive, or dangerous methods with safer ones, and enabled agriculture, resource gathering and mining, and transportation to be carried out faster than ever. Because so much time was saved and so many people's lives transformed for the better, this freed up enough for millions of people to innovate and devise ways to improve their situation, rather than merely take care of their survival.

4. *The spread of computing.* The fourth revolution is known as the digital era. Virtually all the communications technology we take for granted today came with the painstaking invention of the personal computer, the Internet, and more.

Digitalisation means even more difficult and dangerous jobs can be done by computerised robots, 'employees' which do not need to be paid or take days off. Activities that once needed us to go out can now be carried out quickly and easily online, with just a few clicks of the mouse—and now, taps on a smartphone.

A fifth revolution is happening today, building on the technology developed in the previous eras and applying it to novel uses that were unheard of only a few short years ago. 'In short, it is the idea of smart factories in which machines are augmented with web connectivity and connected to a system that can visualize the entire production chain and make decisions on its own', writes Bernard Marr of *Forbes*. New technologies will 'combine the physical, digital and biological worlds'.[3] All the past revolutions have changed the nature of jobs and how we add value to society—and the fifth will be no exception.

After all, why pay dozens of farmhands to harvest crops when a combine harvester can do it much more efficiently? Why pay factory workers to assemble cars when a robot can do it so much faster, and go without pay, food, or sleep? Why go to the trouble of arranging for deliveries and auto repair when a computer can track your needs and automatically fill them, all without your lifting a finger? AIs are 'farmhands' that can do these jobs today, if we only let them—and they need no food, sleep, or pay to multiply our efforts. All they need is processing power, which is so commonplace today that each of us carries millions of times more computing power in our pockets than was used to send the Apollo astronauts to the moon.

Of course, this simplicity belies an incredibly complex system of computer programmes, diagnostics, and machine

---

[3] Bernard Marr, 'Why Everyone Must Get Ready for the 4th Industrial Revolution,' *Forbes*, 5 April 2016, at https://www.forbes.com/sites/bernardmarr/2016/04/05/why-everyone-must-get-ready-for-4th-industrial-revolution/#7a389e193f90.

intercommunication that needs entire teams of human beings to build, test, and maintain. As the nature of work shifted from labour-intensive to skill-intensive, people needed to shift with it. Out were the indentured servants—and in were the engineers, builders, and coders who kept things running.

'While technology has placed the power in consumers' hands, organisations need to harness the same technology not only to meet the demand they bring, but also to monetise their assets,' says Garry Ng, vice president and head of information security at home-grown gaming hardware firm Razer. 'An organisation does not digitise and innovate to unlock these new values will fade away very quickly.'

AI: The Everyday Revolution

I share in my talks that we need to be very clear on what AIs are and what they do. Despite all the hype, I prefer a simple definition of an AI—a computer programme that learns to solve problems and reach a pre-existing goal given by its human programmers. For all intents and purposes, it is a computer programme designed as a tool to replace some aspect of human thought. The field known as AI, then, is a discipline encompassing computer science, data set processing, and machine learning. It is concerned with the creation of algorithms that learn better and better ways to use input data to solve the problems that they are given.

This is the sum total of all AIs do, in the same way our muscles merely take electrical signals from our brains, and contract or relax accordingly. What we accomplish with this ability is up to us, and whether they're good or bad things is independent of their basic function, subject to value judgements from the outside. Our muscles are but tools for our brains, and so are computer programmes like AIs.

For instance, when Boston Dynamics's robot 'dog' walks around and learns how to navigate stairs, right itself when it falls and avoid obstacles, it's not actually doing so because it 'knows' it must do these things or that this is what its creators expect of it. What is happening is that it's collecting data from sensors around its body, interpreting that data according to its programming, and executing a response to what the data says it is, for example, a forward step, an obstacle, or a fall over. In the process, it 'learns' by adapting that response so it does better next time.

What happens as it responds forms another data set, which it interprets once more—and so on and on. All this can happen millions of times a second, in a simulacrum of the biological nervous system. When we watch this happening in the awe and wonder it deserves, the programme merely does what its human creators have told it to do, albeit learning 'how' to do it in better, more efficient ways. It doesn't care about our responses.

For simplicity, I divide such programmes into *narrow, general*, and *super* AIs. I call today's (and very probably tomorrow's) AIs *narrow* because they can only do one task but do it extremely capably and efficiently. Think of the Deep Blue supercomputer that beat world chess champion Garry Kasparov in the 1990s (a precursor to today's far more formidable chess engines) or Apple's iPhone assistant, Siri. Other narrow AIs include the software that powers smart homes, or Tesla and other companies' self-driving cars.

The second category is *general AI*, which mimics a human's or animal's ability to perform multiple tasks by applying knowledge in multiple contexts. They still cannot fully imitate a human intelligence, although some definitely come close and might even pass a Turing test—that is, it can answer general questions and pass as human to an outside observer who does not yet know they are speaking to a computer.

Such AIs have not yet been realised, but some narrow ones indeed approach this level of capability. One example is the interaction programme GPT-3, which can write articles about itself and mimic a human writer. For instance, for several weeks, a copy posed as a human on Reddit and gave the following advice to people with suicidal thoughts:

> I think the thing that helped me most was probably my parents. I had a very good relationship with them and they were always there to support me no matter what happened. There have been numerous times in my life where I felt like killing myself but because of them, I never did it.[4]

Perhaps one of the most promising fields for general AIs is writing and content creation. I don't see it replacing human writers, film crews, and actors, but they do save us much of the tedium that goes into those roles by simulating how a perceptive human might approach it. If I were writing a blog post, for

---

4   Quoted in Will Douglas Heaven, 'A GPT-3 bot posted comments on Reddit for a week and no one noticed,' *MIT Technology Review*, 8 October 2020, at https://www.technologyreview.com/2020/10/08/1009845/a-gpt-3-bot-posted-comments-on-reddit-for-a-week-and-no-one-noticed. The post had been upvoted over 150 times by that writing.

instance, I might turn to a GPT-3 service to outline it for me, insert keywords, and produce a rough draft. Then with the time spent drafting saved, I can sit at my computer and polish it to match my research, goals, and writing style. The first draft only needs the key points—it doesn't have to make sense yet because this is something a human writer is much more capable of inputting into the work.

Finally, there are fictional *super* AIs, which is to present-day AI technology what Superman is to ordinary people. Think HAL from Arthur C. Clarke's *Space Odyssey* stories, JARVIS in Marvel's *Iron Man* comics, or *Terminator*'s antagonist Skynet. That said, fears that some AI may decide to wipe out humanity in the future are more informed by movies than real life. Perhaps what *Jaws* did for sharks, *Terminator* and other such movies have done for AIs.

How likely machine intelligences are to reach this level is beyond the scope of this book, but who's to say they will *never* become real? After all, an inexpensive device in your pocket that gave you access to instant communication, high-quality video and vast swathes of human knowledge would have been firmly in the realm of science fiction in our grandparents' day.[5]

AIs have already entered—some might say infiltrated—every part of life. 'Pandora's Box is open, and AI is no longer the far axis of science fiction,' Serene Keng, managing director of channels and alliances for Asia Pacific and Japan at graph

---

5 Think twice before dismissing them as out of touch, though. As former US president Ronald Reagan pointed out when he was governor of California in the early 1980s: 'We didn't have those things back in our day. We invented them.'

analytics firm TigerGraph, points out. 'But rather, it will be a fixture in our lives, whether we are conscious of it or not.'

Think about the things you do not do just at work but in your daily routine. At home, you might fire up Steam to play games or watch YouTube videos. When you shop in this age of COVID-19, most malls would have a thermal camera set up to read your body temperature as you enter. When you drive or book a transportation service like Uber or Grab, a sensor installed in your car or the relevant smartphone app logs where you go, when you arrive and leave, and how quickly you travel between locations.

In between all these, chances are you'll be doing Internet searches, reading, and playing games on an Apple or Android smartphone. If you interact with a computer in any way, you're effectively giving said computer's owners a little piece of information about yourself, down to the very words you use.

Done enough times over the course of a year, you'll have left an indelible electronic 'signature' that reflects on how healthy you are, your personality, what you're interested in, where you travel and why, and more. Anyone who can see the totality of that data will know who you are, perhaps better than you know yourself!

People change as time goes by, and with you, so will the data you generate. A programme that 'sees' those changes would understand all these versions of you, effectively tracking the sweep of your life from cradle to grave. Multiply that by the billions of people being so tracked. That's one very big data set!

I've written all this to answer three questions. The first is a *what*—what does the explosion of AI technology mean for life in the future, particularly in the next two generations? While science fiction is great fun and a way of asking deep philosophical questions, the truth is often far more mundane.

Perhaps one of the most enduring movies out there is *2001: A Space Odyssey*, released in 1968 and featuring permanent space stations and a thriving moon colony served by Pan Am flights as regularly as far-off cities are reached by airlines today. When the real-life 2001 came around, the moon had not been visited in decades, rudimentary AIs were around but were a far cry from the movie's HAL character, and the actual airline Pan Am no longer existed. Another film that's aged badly is *Back to the Future*, made in the 1980s but set in 2015. When that year actually arrived, the film's iconic hoverboards were nowhere in sight.[6]

These movies may be fictional, but they do illustrate that human beings, even those we rightly remember as great visionaries, aren't very good at predicting the future. As IBM president Thomas Watson said in 1943, 'I think there is a world market for maybe five computers.' Some time later, Microsoft founder Bill Gates would declare, '640K of memory should be enough for everyone.'[7]

---

6   Science fiction does get some things right, as we shall see in this book. For instance, an eerily prophetic *Archie* comic of 1997, set in the then-far-off year of 2021, does predict home-based, online learning replacing going to school in person. This became reality for millions of students, thanks to the COVID-19 pandemic.

7   Today, a desktop or laptop computer you can buy in a store comes with 8, 16 or 32 gigabytes of memory—hundreds of thousands of times Gates's early estimate.

Advancing in the Right Direction

We can, however, make educated guesses if we limit our scope. This brings us to the *how* question—in what ways will automation and AI transform our lives in key areas? Much more could be written, but I want to focus on four:

- *Climate change.* World weather and climate conditions are changing, sometimes rapidly—leaving society and organisms around the planet to adapt or perish. How will AI augment our ability to mitigate the damage these changes will cause, lower the inherent fragility in our civilisation today, and make ourselves more adaptable?
- *Silver populations.* This is a particularly pressing issue in the developed world, as more and more people grow busier or live lifestyles that encourage having fewer children. Our populations are replacing themselves less and less, but the flip side is that medicine is both extending our lives and keeping them productive for many more years. What advances can we expect as a result, and how long can we and our children expect to live?
- *Epidemic and pandemic control.* I'm writing this amid the COVID-19 pandemic, and the way it's being managed compared to plagues of the past is a fascinating study in itself. Vaccines are being distributed and refined, and governments are moderating their responses accordingly—and in the right hands, AI has been a powerful tool in directing policy and planning where they will be most helpful. How will pandemics of the future be handled?
- *Household appliances and the Internet of Things.* The concept of IoT is making daily living 'smarter'

by installing ambient décor and appliances that automatically react to our presence and wishes in our homes, and changing the way people commute. For instance, a home AI might be able to operate your refrigerator, microwave, washing machine, and more, seamlessly ordering more supplies when you're running low, adjusting the air conditioning for you and more. While it may seem mundane, it is this field that has the most potential to revolutionise the way we live and work. Great strides have already been made through AIs like Amazon's Alex, Google's Assistant, and Microsoft's Cortana. On the information front, AIs have already taken humans' places as coaches, artists, and influencers. How will these fields change as a result?

The saying goes that fire is a good servant and a bad master, and the same is true of AI. As CEO of AI and metaverse-centric enterprise, Imperium, I've grown deeply acquainted with the use of AI in everything from the metaverse, cybersecurity to training and hope I can shine a light into adopting these new trends with wisdom, caution, and adaptability. AI is there to do new things that helps live better lives, not simply do the same things more efficiently.

'With today's computing power and abundance of data, we can decide how to fully reap the benefits of AI technologies to build sustainable businesses and improve the environment,' says Hanyong Low, head of technology solutions for group IT at the Kuok Singapore Limited Group. 'Be it healthcare, security, energy conservation, agriculture or any other field, AI is creating many more new jobs than it replaces. Embrace it!'

Thus, the third question we're answering is how we can best let AI help us so that we can harness its potential without letting its vulnerabilities open up new ways for us to be harmed. The benefits are there, but we must also keep ourselves informed about weaknesses.

For instance, a self-driving car that will never get into an accident sounds wonderful, but what if hackers take control of one—or even an entire fleet of them? (This has actually happened in testing, and if the hacks surrounding blockchains and cryptocurrencies are any indication, it will be an arms race, and a recurring problem, well into the future.) This is to say nothing of attacks on the larger infrastructure, like the production and reliable delivery of power and water. After all, even the most sophisticated drones are useless without a way to keep them charged and flying!

The Next, Next Lap

Next, we'll take on a case study of Singapore, my home country, and how it has embraced the advantages that AI brings across various sectors of society—giving policymakers and the private sector alike an unparalleled look at the effect new actions and rulings have. I'll be drawing observations and lessons from its balance between aggression to create and seize new opportunities, and the caution needed to examine their consequences before acting.

Finally, I'll close with a look at human consciousness and what AI will do to strengthen our capabilities as leaders and thinkers. There's something to our thought and cognition that AI will probably never approach, and for it to continue to augment our capabilities in the long term, we'll need a higher view of our fellow humans and their worth to us, the world, and the universe. That will be how we ride the wave of the future and keep ourselves from falling off—to the benefit and not the cost of everyone else.

Thanks for coming along with me on this ride into the future. I hope your seat belt is secured!

ONE

On the Shoulders of Giants

> Cars that drive themselves were invented ages ago. They're called taxis.
>
> James May, *Top Gear*,
> season 13, episode 5

In the late evening of Saturday, 17 April 2021, two men in Texas get into the front passenger and rear seats of a 2019 Tesla Model S. According to witnesses later, neither of them is sitting behind its wheel—if this is true, they have entrusted the vehicle's Autopilot AI with not only their journey but also their very lives. It is likely they have done this before, and for thousands of drivers in the past, the system has delivered. With little or no input save for a starting point and a destination, Autopilot takes control of the vehicle and delivers its occupants wherever they want to go.

Of course, there's no way to eliminate accidents. But Tesla claims that they are less likely to happen than without driver assistance features, and one Autopilot-engaged accident is claimed to happen only every 4.19 million miles driven as of March 2021. Without Autopilot but with active safety features on, an accident is said to happen every 2.05 million miles driven. Without both, accident rates increase to one every 978,000 miles driven.

According to the US National Highway Traffic Safety Administration (NHTSA), Tesla says, 'In the United States there is an automobile crash every 484,000 miles.' By these metrics, Tesla believes its technology is making cars safer by cutting the probability of an accident in half, at the very least.[8]

Sadly, those two men are about to join the ranks of a new but growing group—the victims of self-driving car failures. Sometime during their journey, their Tesla veers off the road and slams into a tree. Despite the best safety features available to the company, the vehicle bursts into flames. Fuelled by the car's lithium-ion battery, the ensuing blaze will take four hours to extinguish.[9]

As with any tragedy in our 24/7 news cycle, the opposing camps have their talking points and narratives ready, almost before the flames are out and the bodies pulled from the wreck. Tesla CEO Elon Musk denies the Autopilot system is at fault or that it was even turned on at the time of the crash. Authorities on the

---

8   Tesla, 'Vehicle Safety Report, Q1 2021', at https://www.tesla.com/en_SG/VehicleSafetyReport.
9   Charisse Jones, 'Tesla Model S crashes in Texas, leaving two dead and sparking a blaze that lasts hours', *USA Today*, 18 April 2021, at https://www.usatoday.com/story/money/2021/04/18/tesla-model-s-car-crash-leaves-two-dead-texas-and-ignites-blaze/7276828002.

ground, however, maintain that the driver's seat was empty.[10] We may know the truth by the time you read this.

The bottom line is that even if the cars themselves are safe and the numbers bear this out, a fully autonomous vehicle is still many years away, and it is a dangerous mistake to behave as if this were the case, whatever its marketing may tell you. The cause of the most accidents still hasn't changed—as the old saying goes, it's still the nut behind the steering wheel.

Musk himself has been accused of selling a pipe dream that cars can drive themselves, 'even though in the fine print Tesla says they're not ready.'[11] Indeed, this is only the latest (as of this writing) incident involving a Tesla and possibly its Autopilot feature—among many other fatalities in dozens of incidents was former US Navy SEAL Joshua Brown, who was killed when his Tesla collided with a truck in May 2016. Ironically, just a month before the fatal crash, Brown had praised Autopilot for swerving out of the way of another truck and avoiding a collision.[12] It must be noted, however, that these are only a few dozen cases among the thousands, eventually millions, of driver-assisted vehicles on the roads.

---

10 David Shepardson and Hyunjoo Jin, 'Texas police to demand Tesla crash data as Musk denies Autopilot use', *Reuters*, 20 April 2021, at https://www.reuters.com/business/autos-transportation/us-probes-fatal-tesla-crash-believed-be-driverless-2021-04-19.
11 Tom Krisher, 'Scrutiny of fiery Tesla crash that killed 2 in The Woodlands a sign that regulation may be coming', *ABC13*, 21 April 2021, at https://abc13.com/tesla-crash-autopilot-elon-musk-model-s/10531109.
12 Mary Bowerman, 'Witness: Driver watched "Harry Potter" as self-driving car crashed', *USA Today*, 1 July 2016, at https://www.usatoday.com/story/tech/nation-now/2016/07/01/navy-seal-vet-killed-using-teslas-autopilot-posted-close-call-video-month-ago/86592458.

Perhaps the best way to see Autopilot and its optional augmentation Full Self-Driving isn't as a driver replacement but an advanced, albeit still developing, tool to assist human operators facing an inherent level of danger. After all, driving a car may be more convenient for us, but conditions can change in an instant, and as such, the activity requires our full concentration, for the sake of ourselves and others. As such, a vehicle should never be considered fully autonomous.

As Tom Krisher of the Associated Press has noted:

> Tesla's system requires drivers to place their hands on the steering wheel. But federal investigators have found that this system lets drivers zone out for too long.
>
> Tesla plans to use the same cameras and radar sensors, though with a more powerful computer, in its fully self-driving vehicles. Critics question whether those cars will be able to drive themselves safely without putting other motorists in danger.[13]

Tesla isn't the only player in the autonomous vehicle market, but thanks to Musk's profile, it is by far the best known. But this issue does raise a fascinating question, one that should give us pause as we consider the role that AI augmentation will play in the future. We may be deciding what AIs do, but what happens when, whether by accident or design, they decide what *we* do?

---

13 Tom Krisher, '3 crashes, 3 deaths raise questions about Tesla's Autopilot', *AP News*, 3 January 2020, at https://apnews.com/article/ca5e62255bb87bf1b151f9bf075aaadf.

Even before we consider the consequences and ethics of AI-powered vehicles, there is already a profound psychological effect on drivers themselves. Because they feel that the AI can do all the work, they take their eyes off the road, and if their hands are on the wheel, they aren't actively looking out for dangers. It is believed that in Brown's collision, and in another in North Carolina in 2020, both Tesla drivers were watching movies at the time.[14] Without an 'autopilot', how many drivers would dare try such a thing?

Where the Money Goes

Management professor Joseph Fuller has pointed out:

> Firms now use AI to manage sourcing of materials and products from suppliers and to integrate vast troves of information to aid in strategic decision-making, and because of its capacity to process data so quickly, AI tools are helping to minimize time in the pricey trial-and-error of product development—a critical advance for an industry like pharmaceuticals, where it costs $1 billion to bring a new pill to market…[15]

---

14 Tesla itself, however, disputed this. The screens that the car provides to the driver are incapable of playing movies, but nothing prevents the driver from watching it on an external device, such as a smartphone or laptop. Regardless of the equipment used, the lesson remains the same—don't be distracted or impaired while you drive!

15 Christina Pazzanese, 'Great promise but potential for peril', *The Harvard Gazette*, 26 October 2020, at https://news.harvard.edu/gazette/story/2020/10/ethical-concerns-mount-as-ai-takes-bigger-decision-making-role.

This reveals several important factors behind AI augmentation of our abilities, which will grow ever more crucial as we entrust AIs not only with our purchases and preferences but also with our lives and well-being:

1. *How do we handle the rise in unemployment and the job shifts that would result?* AI won't *reduce* the number of jobs available, but it will change around which industries are in favour and which are not. After all, the rise of the automobile meant the end of the horse and buggy one, in the same way that printing presses put scribes out of a job. Perhaps the most vivid example comes from warfare, where the once-feared cavalry was made obsolete by the march of firearms technology.

   Drivers face being phased out as AI cars, trucks, and buses become more affordable and more customers choose them over driving themselves or conventional public transport. As Adam Hayes of *Investopedia* warns:

   > Taxi and delivery drivers account for 370,400 jobs, and more than 680,000 Americans are employed as bus drivers. Taken together, that represents a potential loss of more than 2.9 million jobs—which is the more than number of jobs lost during 2008 due to the Great Recession.

Add in delivery and light truck drivers (1.5 million) and the total number of potential jobs lost grows to a staggering 4.5 million.[16]

'One interesting possible consequence is that after a few generations, very few people will even know how to drive a car anymore,' Hayes notes.[17] I don't see that ability going away any time soon, but it may become a specialised skill—the way emergency first response, car maintenance, or computer technical support are today.

That said, AI assistance does enable a single team member to keep track of and resolve issues faster, allowing them to handle a greater caseload. Those job holders who *do* remain will probably find themselves in greater demand. Consider a customer service department where many routine queries are handled by an AI, with only exceptional cases being handled by a human employee.

We've seen great changes in employment thanks to the COVID-19 pandemic, with mass layoffs and resignations in the United States and the rest of the world. With more people working from home or laid off due to the changes wrought by government mandates, they are more able to think about their purpose and job scopes and consider the deeper meaning of their work. As they have before and they will again, they are rejecting low-paying manual work open to abuse

---

16 Adam Hayes, 'The Unintended Consequences of Self-Driving Cars', *Investopedia*, 15 May 2021, at https://www.investopedia.com/articles/investing/090215/unintended-consequences-selfdriving-cars.asp.
17 Ibid.

and exposure to the elements, especially in the service industry.

This doesn't mean jobs are insufficient or that they're being replaced. Rather, it's a signal that human workers need to move up the value chain, leaving manual, low-value work to automation software and AI-powered robots.[18]

At *Forbes*, Steve Andriole points out that this great resignation is an 'automation stimulus' as human labour is replaced by lower-cost machinery whenever possible:

> Automation has only begun, and as more and more employees call it quits automated workers may take their place faster than we think. The incentives are clear. Why wouldn't Uber want to eliminate their biggest headache—drivers—with autonomous vehicles? Why wouldn't all companies want to deploy 'workers' that work 24/7, never need vacations, never join unions and never get sick (from viruses, at least)?[19]

---

18 It is most fitting that the Czech word for *drudgery* or *forced labour* is *robota*, from which we first derived the English word *robot*.
19 Steve Andriole, 'The Great Resignation Will Only Accelerate Automation. Employers Will Retaliate With Machines', *Forbes*, 16 November 2021, at https://www.forbes.com/sites/steveandriole/2021/11/16/the-great-resignation-will-only-accelerate-automation--employers-will-retaliate-with-machines.

2. *How will ancillary work change as a result?* Today, an entire industry of rest stops caters to long-distance travellers around the world, providing them with rest, meals, and toilet breaks. What happens to it as driverless cars grow more and more common is an open question, but Colin Priest, data scientist and global lead on AI governance at DataRobot, believes there will be *more* demand for such ancillary workers, not less.

'Humans are inherently social, and behavioural science research shows that when customers see human effort in a product or service, they appreciate it more and believe it to be higher quality,' he says. For instance, travellers who have been isolated or had less human company will be more appreciative of seeing cashiers, waitstaff, and other people at rest stops.

Perhaps one AI-enhanced sector where this is noticeable is customer service, in which customers feel the best when assisted by a human employee with the ability to solve their problem. Where an AI will need complex, cutting-edge programming to parse their words, filter out sarcasm, and finally guess at a customer's true meaning, a well-trained employee instinctively does this in seconds.

It's a reminder that while humans lack an AI's memory and data-crunching capability, computers lack the ability to 'read' people and empathise with them. This is what makes them our tools, not our masters.

3. *Security is already a crucial enough issue.* Can we be assured hackers will not take over? This has happened

in tests, and when the hardware and software become advanced enough to take complete control of a vehicle, the complexities of the code may leave open doors for someone else to seize it—and even crash it, if they so choose.

Whatever the use, an AI can only learn and do what it is told, and sometimes this can go wrong as Tencent researchers once showed when they tricked a driverless car into wrongly changing lanes simply by placing stickers on the road. Despite its computing power, it's not capable of the intuition and gut-level, instinctive thought that a human is.

The danger only increases if they are networked together and share information. Even one vehicle becomes a terrible danger in the hands of a malicious or incapacitated operator, to say nothing of an entire fleet of them. If a hacker (who can be located anywhere in the world) does it, they could wreak untold havoc at will.

This does underscore the growing need for cybersecurity, and the related development of prevention and diagnostic tools. Like it or not, the industry is in an arms race every bit as real as the US–Soviet one of the Cold War.

4. *How will manufacturing and insurance move forward as the risks change?* Fewer and fewer people are expected to own cars as the hassle of maintaining your own is replaced by the ability to summon an automated 'taxi' to take you wherever you wish. That will probably result in fewer *cars* since those that do exist will be owned by

their providers and shared among a pool of customers. The result would be manufacturers being forced to lay off workers, downsize, and accept the loss of much of their revenue.

Insurance will also be affected. Because insurance firms' revenue rises with the risks customers are expected to face, reducing the risks of driving and making people safer would drive down the prices of insurance policies. The industry will have to change the way it handles risk assessment, and it's worth watching in the future.

5. *How will liability change as AIs grow more sophisticated and approach some form of self-awareness?* At present, Tesla's Autopilot is still dependent on human programming and input—and the company itself still faces scrutiny if reliance on its software leads to accidents. It is, after all, still just a narrow AI that isn't self-aware, truly autonomous, or capable of setting its own goals. At present, there is no such thing as a 'rogue AI' because whether the results line up with our intentions or not, all software simply does what human programmers have told it to do.

But as we entrust more and more to it, and its capabilities grow to the point of directing its own growth and conversing with drivers, more such questions will need to be answered.

It's worth noting, as I did with then-Daimler Southeast Asia CEO Franco Chiam, that aviation has had a far greater level of automation and computer assistance for generations—but driving remains strongly resistant to it. Perhaps it's because we do it far more often, see obstacles and difficulties around us, and feel more of a need to be in control. On the other hand, airliners operate in a vastly different environment, yet one that is fairly predictable. Besides, despite all the automation we entrust our lives and safety to, we're more inclined to trust a human captain and co-pilot at the controls.[20]

The Leapfrog Effect

Wherever they're deployed, AIs are gathering data and 'learning' to interpret it at a phenomenal pace. It's helped us solve some of the world's most pressing problems, from protein folding to the vagaries of banking and finance, and the interpretation of medical images. While it's not always translated into concrete changes in products and services (yet), what *is* happening is that our achievement has skyrocketed with it. We live better, work better, and collaborate better, making more efficient use of limited resources.

Consider the case of the Ant Group, Chinese billionaire Jack Ma's investment firm. Because fintech and AI integration have been part of its operations from the start, it's proven that it can leapfrog more established institutions like Bank of America (BOA) and Goldman Sachs. Despite its smaller size, Ant has grown to serve nearly half a billion users annually, compared to BOA's 66 million. All those users are served by Ant Group's 16,660 employees, while BOA has over 140,000.

Simply put, AI-assisted institutions empower the few to serve the many, with human operators leveraging advanced computer

---

20 For more, see: Tony Tan, 'Driverless Cars: Is It Happening?' *YouTube*, 26 February 2021, at https://www.youtube.com/watch?v=50jxa02QrKMs.

software that does in seconds what would take a human centuries. By understanding patterns in the data, it can point out what needs to be done—and then its operators can roll out services right at the time they are needed and use cloud computing to seamlessly connect users all over the world. Those institutions that build AI assistance into their operations from the start can, if they play their cards right, catch up with and even overtake those that are larger and more established, but slower to change.

This has happened in the background, but it's slowly grown more accepted because it gives *us* more control over our portfolios. Few of us can study the performance of every listed company and pick the best stocks based on our own risk appetites, but an investment software merely asks us to make a choice: 'Do you want your investment portfolio to be mild, moderate, or aggressive?'

You, or your broker, can simply let the software run according to the rules each setting gives it. That allows each broker to serve more clients, reducing the amount each one pays in commission. Everyone wins! But again, because computer software only does what it's told, there have been near misses, and there are many opportunities for threat actors to game them, as well as mistakes what it is taught. All these have real-world consequences.

A Word of Caution

In a number of cases, AIs have fallen so far short of their expectations that they have had to be rethought or replaced entirely. This has been the case even if it was from a trusted provider. For instance, in June 2021, Nicole Wetsmen of *The*

Verge covered an algorithm designed to detect signs of sepsis infection being deployed in hospitals around the United States. It was from Epic Systems, the largest electronic health records company in the country—but in a Michigan hospital, it missed 1,700 sepsis cases, out of 2,552 hospitalised patients who eventually developed the life-threatening infection.

'The analysis also found a high rate of false positives: when an alert went off for a patient, there was only a 12 percent chance that the patient actually would develop sepsis,' Wetsmen wrote. It turned out that:

> The algorithm used information on bills for sepsis to define which patients had sepsis. That means it's catching cases where the doctor already thinks there's an issue. It's also not the measure of sepsis that researchers would ordinarily use.

Such tools are 'only as good as the data they're developed with, and they should be subject to outside evaluation'.[21] Other once-promising uses of AI have petered out due to the technology not being sufficiently developed or proving too risky in the real world or not cost-effective enough. (For instance, shopping malls are among the most visible and riskiest places AI-controlled robots can be deployed, and YouTube has several clips of robots guiding themselves to escalators and tumbling down them, or ending up in display fountains.)

---

21 Nicole Wetsman, 'A hospital algorithm designed to predict a deadly condition misses most cases', *The Verge*, 22 June 2021, at https://www.theverge.com/2021/6/22/22545044/algorithm-hospital-sepsis-epic-prediction.

Other companies have decided their resources are better used elsewhere; even ride-sharing giant Uber closed its AI labs and incubator in May 2020 and sold its driverless car division to a start-up called Aurora.[22]

I hope you'll see the spirit of this book going forward. I want to bring the future into the present and reveal it for what it truly is—not fuzzy-headed or over-optimistic scenarios with more to do with science fiction than science fact, or doom-and-gloom jeremiads that scare people into believing that, say, climate change will turn the entire planet into a hell on earth. Even if this technology is not yet ready for the road in densely populated urban areas, we can be confident that it is well on its way to further improvement and mass adoption in the coming decade.

Like every other advance I'll talk about later, autonomous vehicles, financial tech that does the hard decision-making for you, and other novel uses of AI are merely tools. It will be the mindset we apply to them that makes them good or bad. After all, the difference between a car cruising on the road and one ploughing into a crowd of pedestrians is only a few metres (or a few millilitres of alcohol in the driver's system), but it's also the difference between life and death.

---

22 '2020 in Review: 10 AI Failures', *Synced*, 1 January 2021, at https://syncedreview.com/2021/01/01/2020-in-review-10-ai-failures.

TWO

You Can't Spell Gaia Without 'AI'

> I believe real progress can be made on most issues in the natural world by 2030, let alone by 2070, if we all urgently come together. To achieve all that the world needs to do, we need governments to treat the climate crisis as seriously as an invasion on their countries and join together to whip it.
>
> Richard Branson

Thailand's beaches are beautiful and popular with holidaymakers, and in late December of 2004, a ten-year-old English girl named Tilly Smith was at Mai Khao Beach in Phuket when she noticed something strange. The bright-blue water began to fizz as if it had turned to clear soda. That wasn't all that got Tilly's attention—the water 'seemed to be rolling farther up the beach than it had a few minutes ago'.[23]

None of the other holidaymakers on the beach paid any notice, not even Tilly's own mother. But the alert girl noticed yet a third oddity, 'a log spinning in circles in the sea'.

In a logical feat that AI still struggles to replicate, all three disparate signs came together in Tilly's head. Recalling a lesson on how tsunamis formed that she had been taught in geography class just weeks before, she proceeded to do what hundreds of seismic sensors placed throughout Southeast Asia could not. 'Tsunami! There's going to be a tsunami!' she shouted. 'We have to run!'[24]

Tilly's father alerted a Japanese security guard, who had heard about the triggering earthquake in the Indian Ocean, but not that the resulting waves would reach as far east as Thailand. He raised the alarm, and everyone on the beach began to run for their lives. They were just in time, and as water began to flood their hotel, Tilly urged everyone to stay on the high ground.

Only later did they learn what had happened. What would become known as the Boxing Day Tsunami wreaked havoc across South Asia, killing nearly 230,000 people in fourteen countries—one of recorded history's worst natural disasters. But thanks to one girl's quick thinking, and a most timely lesson

---

23 'Tilly Spots a Tsunami', *The Red Cross*, (no date), at https://www.redcross.org/content/dam/redcross/atg/NHQ_PDFs/Tsunami_Activity.pdf.
24 Shabdita Pareek, 'Here's How A 10-Year-Old Girl Saved Everyone On A Thai Beach During The 2004 Tsunami', *Scoopwhoop*, 26 December 2015, at https://www.scoopwhoop.com/tilly-smith-tsunami-2004.

back at school in England, some 100 people at Mai Khao Beach were not among them.

Tilly herself would be feted for her courageous actions and return to the same beach on the tenth anniversary of the disaster. She has certainly more than earned a relaxing holiday there!

That said, who knows how many more lives could have been saved with better advance warning of the tsunami? To replicate what Tilly did at scale, an autonomous network of sensors would have to be deployed in oceans around the world, in a cost-effective way—with the ability to sort the vast amount of data they would collect, interpret it in a way that would spot the signs of a tsunami, and warn the authorities.

In the wake of the 2004 tsunami, Indonesia installed a multimillion-dollar tsunami warning system, utilising twenty-two buoys. When the system was put to the test by an earthquake off Indonesia's west coast, none of the buoys worked—meaning officials knew it had occurred but lacked any information how severe it was. 'Authorities issued a tsunami warning, but struggled to acquire the data needed to rescind it quickly, waiting nearly three hours until it was clear that massive waves hadn't formed,' notes the *Wall Street Journal*. 'Hundreds of thousands of people fled, waiting for hours to see whether a monster wave would come ashore. It never did.'[25]

In 2017, Hiroko Sugioka of Kobe University began tests of a satellite-uplinked, wave- and solar-powered robot to sail the seas

---

25   Sara Schonhardt, 'I Made Sentana and Anita Rachman, Earthquake Exposes Gaps in Indonesia's Tsunami-Warning System', *The Wall Street Journal*, 3 March 2016, at https://www.wsj.com/articles/earthquake-exposes-gaps-in-indonesias-tsunami-warning-system-1456997452.

without needing to refuel for a year. By connecting with sensors on the ocean floor that pick up the telltale changes in water pressure and magnetic field, the cost-effective robot can issue an alert in just a few minutes—giving tsunami control centres on land the opportunity to issue a warning, and people enough time to reach higher ground.[26]

Disasters will happen, but tracking the movement and shifts of the earth beneath our feet has many more implications for solving, or at least mitigating, the problems that the wider issue of climate change will bring.

Doing More with Less

Sea levels are definitely rising in many places, and low-lying countries are more vulnerable to this. Climate change has historically threatened those least able to adapt to it, like farmers and islanders; the changing weather patterns destroy crops or make them unsustainable to grow in the first place, and island dwellers must move away—or risk being swallowed by the sea.

One way to mitigate the issue is to reduce our reliance on carbon, which means more efficient, environmentally friendly ways have to be found. The good news is that our green energy is now adopting faster and faster, and it's growing more efficient and cheaper every year.

---

26 Michael Reilly, 'This Robot Will Sail for Months on the Lookout for a Tsunami', *MIT Technology Review*, 30 January 2017, at https://www.technologyreview.com/2017/01/30/154301/this-robot-will-sail-for-months-on-the-lookout-for-a-tsunami.

Harnessing more green energy isn't sufficient. It also needs to be optimised, and as in the beginning, it will have some way to go before it catches up to more conventional means of energy production. It will need to be better-used; despite its renewable nature, we're limited in how much we can collect. In other words, it is still a scarce resource, bound by economic constraints. Much needs to be done to improve the efficiency of wind and solar, and the collection of alternative sources of energy.

And until those become viable replacements for fossil fuels, other forms of energy such as natural gas and geothermal must fill the gap. Nuclear plants reliably generate clean energy but create hazardous waste products and can cause disastrous consequences in the event of an accident.

One of the most ambitious goals the developed world has set is to achieve net zero carbon dioxide emission by the mid-twenty-first century. But as the *MIT Technology Review* notes, 'nearly half the cuts will have to come from technologies that are only in early stages today'.[27]

These include drastically increasing the capacity and efficiency of batteries, finding biofuels for aircraft and vehicles that can replace the status quo, and deploying machines that remove carbon dioxide directly from the air. The technologies needed to do this are not yet available or cost-effective enough to deploy at scale:

---

[27] James Temple, 'Half of the world's emissions cuts will require tech that isn't commercially available', *MIT Technology Review*, 18 May 2021, at https://www.technologyreview.com/2021/05/18/1025027/half-of-emissions-cuts-require-tech-innovation-climate-change-net-zero.

By 2030, the world must add more than 1,000 gigawatts of wind and solar power capacity annually, which is just shy of the total electricity system in the US today. Electric passenger vehicles need to reach 60% of new sales by 2030, while half of heavy trucks purchased must be EVs by 2035. And by 2045, half of global heat demand must be met with heat pumps, which can run on clean electricity.[28]

One of the great challenges going forward is how to do more with less, and as always, AI's ability to sort, scan, and interpret enormous data sets is a key to the automation we need.

I know I pointed out Tesla's continued safety and perception issues earlier, but it's now time to see how fully electric vehicles can make a difference. The basic principles of automotive engineering have been around for more than a century, but Tesla is one of the first companies to figure out how to make them fully electric, with all the performance and capability one expects from conventional, gas-powered vehicles. Tesla isn't a car company that figures out how to include batteries; it's a battery company that happens to place its products in cars.

But the key lies in market uptake, not government demands. The market is what determines the inventions that get more common and cheaper, and while the climb for fully electric vehicles is uphill, it's by no means impossible.

With a simple charging port that requires no more complication than plugging in your smartphone at night, and with a charging

---

28 Ibid.

time that gets shorter and shorter, electric cars are emerging as a cheaper, more convenient, and environmentally friendlier alternative.

Each charge will also last longer, saving drivers from distance anxiety—the fear that you'll run out of power mid-journey, with no charging port . . . and petrol stations that won't be of any use! It's only a matter of time before the infrastructure to allow longer and safer electric car journeys is in place, just like entire industries sprang up following the proliferation of cars back in the early twentieth century. Where electric cars once needed to be charged for several hours, new batteries have been developed that reduce it to as little as five minutes—depending, of course, on the battery capacity, the power transfer efficiency, and other factors.

But CEO Elon Musk isn't stopping there. He's aiming to colonise space and has plans to put astronauts on Mars. In fact, through his Boring Company and other firms, he's already testing the technologies that will one day make space travel and settlement a reality. His chief rival is Jeff Bezos, himself a founder of space exploration companies and aiming to build a permanent base on the moon.

Whether this eventually happens or not, the technologies they develop will find their way into the marketplace, and like the innovations of the twentieth century, they'll forever change the way we live. The lessons it taught us are many, but chief among them is never to underestimate our potential, both for good and for evil.

That said, there remains much work to be done. Electric car production is another area where environmental friendliness remains to be improved; behind every Tesla vehicle is an enormous manufacturing, mining, and metal-refining chain, and the environmental impact of doing all this continues to be immense. It's still as carbon- and manpower-intensive as it's ever been to dig the required minerals like iron, tin, aluminium, or cobalt out of the earth. For now, we know what *can* be done, but it's an ever-improving journey as we explore what *should* be done.

Used well, AI speeds up research and development by finding trends and patterns, giving us leads and avenues to explore, and improving our ability to collaborate.

Stewardship and Flourishing

If something is crucial enough, the resources will be found to meet the challenge. It lies in getting people to club together for the benefit of all, and the problems posed by a changing climate will be pressing ones that both we and our descendants will face.

Far from being purely a man-made concern, it's been this way for as long as we've lived on earth—the difference is that more recently, we have *both* a more polluted and threatened planet to restore, *and* the information and know-how we need to engineer a better, more stable outcome for ourselves.

Whether our present technologies can make a difference and mitigate the effects of the massive increase in atmospheric carbon intent is still unknown, but the ball is rolling as we move towards finding out and doing what we can.

As with the general tone of this book, I have no wish to dismiss the dangers of climate change or deny it is a problem that could potentially devastate our countries and economies if we let it. That said, neither do I want to exaggerate the danger as many influential thought leaders are wont to do. Doomsaying that climate change will destroy humanity, and talking of 'tipping points' where we only have a few years to save the planet, makes for riveting headlines. But these only serve to raise eco-anxiety and make sober, considered understanding less likely.

So-called panic porn tells us little about the complex issues at hand . . . and much less how to resolve them. The answer isn't to panic and try to drastically re-engineer civilisation but to make adaptations so that we balance both environmental stewardship and the human flourishing needed for it to happen.

As with any large-scale problem, three pressing questions need to be answered:

- *Why are things as bad as they are?* This involves examining large data sets, interpreting what we find out, and finding patterns that warrant further investigation.

- *What would it mean to solve or mitigate the problem?* How much will it cost in resources and the other uses they must be put to? This involves careful planning and analysis of various scenarios and alternative, enabling better, more efficient means of generating energy.

- *What goals need to be achieved before the problem can be considered solved or successfully mitigated?* This needs companies, governments, and individuals to consider their impact on the environment and, as far as possible, reduce carbon footprint and maintain the ecological balance of natural habitats.

There are no silver bullets. Every small effort made towards improving environmental friendliness helps, and it may very well be small drops that make up the ocean that keeps the damage of a changing climate at bay.

1. *Tracking emissions of harmful greenhouse gases.*

   Methane is an even more potent greenhouse gas than carbon dioxide, and this poisonous and odourless gas is a by-product of oil and gas production. Its production typically happens far from population centres and under stringent safety protocols, and exposure can be lethal to humans. 'This main component of natural or fossil gas is a vastly more potent warming agent than CO2 over short timeframes, is accompanied by highly toxic chemicals, and when leaked can be a major health and safety hazard,' writes Deborah Gordon of RMI.[29]

   One of the biggest dangers it poses to human and environmental health is if it leaks from its transfer and storage facilities at any point, and it is crucial that leaks are found and repaired. Companies like GHGSat and BP use AI-assisted satellite imagery and drones to monitor sites and trace emissions, allowing technicians to precisely locate the source of the leaks.

2. *Making energy production more efficient.*

---

29 Deborah Gordon, 'Methane: A Threat to People and Planet', *RMI*, 7 July 2021, at https://rmi.org/methane-a-threat-to-people-and-planet.

The fluctuations of energy demand mean power generation must match it as closely as possible. Wind and solar energy are particularly dependent on weather patterns, and rather than rely on human guesswork, AI is helping to track historical patterns and project the storage and supply needs in the present.

According to the Capgemini Institute, one such plant in the Azores 'improves reliability of the energy grid while aiding larger adoption of renewable power from 15% to 65% and reducing the need for 17,000 liters of diesel per month'.[30]

3. *Optimising routes to travel less and save fuel.*

Delivery firms use AI to analyse traffic conditions and plan the best possible route, reducing both the distance travelled and the number of vehicles needed to deliver the same volume of goods.

4. *Analysing wildfire patterns, allowing better response and life-saving.*

In hot and dry forested regions of the world, forest and brush fires are a constant danger, threatening thousands of people and affecting the livelihoods of millions more. With AI analytics improving data gathering and indicating where fires might begin and how they are likely to spread, officials would be better able to draw

---

[30] Capgemini Research Institute, 'Climate AI: How artificial intelligence can power your climate action strategy', *Capgemini*, December 2020, at https://www.capgemini.com/wp-content/uploads/2020/12/Report-Climate-AI.pdf.

up more effective evacuation and resource distribution plans.

5. *Improving reliability in manufacturing and reducing waste.*

    By using AI to scan the manufacturing process and customer buying patterns, defect and wastage rates can be slashed.

The Capgemini report points out that each benefit—reduced emissions, improved efficiency, reduced wastage and dead weight assets, and saved costs—has improved between 10.9 per cent and 12.9 per cent from 2017 to 2019.[31]

What happens next? As AIs improve, we can expect them to be a valuable tool in making progress towards net zero emissions, but while entire industries are integrating AI capabilities into their overall strategies, the Capgemini analysis points out that 'only a small minority of organizations [have] what it requires to successfully leverage AI solutions for climate action'.

It's also not lost on climate experts that the computer power needed to programme, train, and deploy these AIs are themselves a major driver of climate change. While small compared to an entire organisation's worth of emissions, they do have an impact that must be measured.

This is a rapidly improving field that will only grow cheaper and more widespread, and even if the net zero goal is not reached, it will still remain a powerful impetus for engineering solutions to the issues a changing climate will bring.

---

31 Ibid.

THREE

Gold Solutions for a Silver Population

> Oh, the worst of all tragedies is not to die young, but to live until I am seventy-five and yet not ever truly to have lived.
>
> Martin Luther King Jr

Near the southern coast of Spain in the city of Grenada, there sits a majestic ninth-century fortress known to its Arab renovators as the Alhambra (Arabic for 'the Red Fortress'). First built by the Nasrid emir Mohammed ben Al-Ahmar, it was converted into a palace in 1333 by Sultan Yusuf I. After the reconquest of Spain by Catholicism in 1492, the palace was altered to resemble the style of the Renaissance. The Spanish king Ferdinand and Queen Isabella would hold their royal court there and provide explorer Christopher Columbus with their support in his exploration of the Americas.

Described by Moorish poets as 'a pearl set in emeralds', for its beauty amid the surrounding woods, the Alhambra is a UNESCO World Heritage Site and a popular tourist destination. It has indeed hosted some of the bloodiest fighting of medieval times, and one cannot help but feel the weight of history as they walk its halls.

What does all this have to do with the realities of an ageing population and the role AI has to play? In 2018, a South Korean drama series called *Memories of the Alhambra* dipped into the history of the Red Fortress, imagining it as the basis of a new alternate reality game (ARG). Inspired by the runaway success of 2016's smash-hit ARG *Pokémon Go*, *Memories* envisions players around the world re-enacting the battles fought in and around the ancient palace. Of course, things take a darker turn when players whose characters die in the game begin showing up dead in real life.

It's a thrilling ride that blurs the lines between fantasy and reality. 'The blending of human ingenuity and artificial creativity will raise the art of storytelling to stratospheric levels, as though the audience are the characters themselves,' observes Glen Francis, chief technology officer of Singapore Press Holdings (SPH) and board member of the Accounting and Corporate Regulatory Authority (ACRA). To him, the future of entertainment will be a disruptive metaverse, in which 'consumers will become prosumers, producing in the process of consuming media and enriching other's experiences while enriching their own'.

The in-game murder plot aside, *Memories* is a fantastic look at how technology is shaping up to help us experience things we never could before, whether it's by virtue of being too young, too old, or from the wrong place. Any place in history can be recreated down to the best details available to experts—for instance, anyone can visit a digital version of the Egyptian pharaoh Tutankhamun's tomb, even if the actual, real-life site is closed to everyone except archaeologists.

All of us will grow old. Unless we meet the tragic end of dying young, all of us will have Father Time steal our youth, vitality, and independence, until we come to the end of our days. But what we call decline actually begins much earlier as we age past our prime; it becomes especially clear in our late thirties, and as we enter our forties and fifties, we'll have lost a significant part of the vitality of our youth. When we speak of technology assisting an ageing population, given the time it takes to develop new products, test them, and bring them to market, we're speaking of assisting ourselves.

As a society bears fewer and fewer children, the proportion of elderly residents increases in what has come to be known as a 'silver population'. Because our productivity declines with age, we simply can't provide as well for ourselves and our families in our golden years, and many of us end up having to be cared for by our own children. The result is that fewer young people must provide for more elderly people. According to projections from the United Nations, by 2050, Singapore will have a population of 6.58 million, with those aged sixty-five and older making up nearly half of these.

The dependency ratio will also halve to almost 1:1, with one adult supporting a child or an elderly person. In 2015, there were 100 adults—persons aged twenty to sixty-four years—supporting about 50 children and elderly persons. But by 2050,

100 adults will have to support about 95 children and elderly persons.[32]

But this isn't all bad news. Older people in a developed society will also have more spending money, accrued over a lifetime of prior work. As we age, we naturally want to remain healthier and happier for as long as possible so that we can socialise, pursue our hobbies, and remain independent. An important driver of this is the wish not to be a burden to our children and loved ones, a sad circumstance that is the reality of many.

That said, ageing may be inevitable—but nothing stops us from seeing it in a positive light. As speaker of Parliament and chairman of the PAP's[33] Senior's Group, Tan Chuan-Jin once put it:

> Should we not try to be able to say this—that our best years are ahead of us ... Can we also aspire to become the best place on earth to retire and grow old in? [...] Can we, at the individual level, embrace this and participate fully in a whole effort to make this vision a reality?[34]

---

32 Siau Ming En, 'Elderly to make up almost half of S'pore population by 2050: United Nations', *Today*, 6 December 2017, at https://www.todayonline.com/singapore/elderly-make-almost-half-spore-population-2050-united-nations.
33 The People's Action Party is a major conservative centre-right political party and is one of the three contemporary political parties represented in Parliament, alongside the Worker's Party and Progress Singapore Party.
34 Quoted in Seow Bei Yi, 'See ageing in a positive light, not as a problem: Tan Chuan-Jin', *The Straits Times*, 14 October 2019, at https://www.straitstimes.com/politics/see-ageing-in-a-positive-light-not-as-a-problem-tan-chuan-jin. The PAP is Singapore's People's Action Party, the ruling political party in the country since the 1960s.

In other words, the pressures and necessities that an ageing population brings should not be seen as a problem but a challenge and source of opportunities.

Healthier, and Happier

Thanks to the advances that medical science has made, we're enjoying longer, healthier lives than at any time in the past. It's not enough to think merely of life span because what's the point of living another ten or twenty years if we spend them in pain, sickness, and infirmity? We don't just want to live longer but spend that extra time more productively as well.

It would mean slowing and, if we can, reversing our bodies' natural process of decline. Companies like eSight Eyewear, which manufactures high-tech vision aids for the legally blind and sight-impaired, and producers of AI-augmented hearing aids that isolate and enhance environmental sounds and speech, are at the forefront of doing this. With the aid of AI analysis of MRI brain scans, advanced algorithms can even predict hearing loss and its effects on language development early in life, allowing earlier and easier intervention.[35]

Also, the longer and more productively we live, the more time we'll have, and the more we'll use and consume. That's already another use case of AI in an ageing population—a means of tracking the kinds of products that ageing yet healthier people will buy. For those who do need medical services, there'll be a boom in pharmaceuticals and nursing. AI already works to

---

35 Karl Utermohlen, '4 Applications of Artificial Intelligence for Hearing Loss', *Medium*, 17 May 2018, at https://medium.com/@karl.utermohlen/4-applications-of-artificial-intelligence-for-hearing-loss-64f3e189847e.

help analyse X-rays, CAT scans, MRIs, and other methods of imaging; collates indicators of future serious illness and allows life-saving interventions to be performed sooner rather than later; and saves doctors work by creating customised treatment plans. This AI assistance allows each clinician or team to handle a far higher caseload of patients than before, saving money training and hiring additional personnel.

There'll always be a need for the human touch in medicine, as Tamsin Greulich-Smith points out in GovInsider:

> However, one of the benefits associated with AI may actually be a disadvantage in healthcare. While AI excels at making unbiased, purely logical decisions, it cannot yet appreciate the complex blend of emotional, social, cultural and physical needs of people. If left in the hands of a purely AI-driven doctor, the diagnosis and treatment plan may be correct, but it may not be what the patient will respond best to.
>
> Human doctors understand how to treat people, rather than just diseases.[36]

A second possible use of AI here would be in relationship-building. Because an ageing population will place increased demands on a society's healthcare system, the impact of keeping patients engaged and compliant with their care plans becomes even more urgent.

---

36 Tamsin Greulich-Smith, 'Can artificial intelligence care for the elderly?' *GovInsider*, (no date), at https://govinsider.asia/innovation/artificial-intelligence-ageing-population.

A patient who is sick and recovers by strictly following their treatment regimen taxes the system less; one who does not and remains sick will create a higher caseload and divert medical personnel's attention from other, more serious cases. The key to keeping such patients compliant lies in setting up the right rapport with them, and despite the heavy workload doctors have, this factor cannot be neglected. Greulich-Smith continues:

> They may be confused about what to do or how to do it; they may feel better and hence decide to discontinue their medicines. They may even forget or lose motivation. This is particularly true for seniors living alone. When you add social isolation into the mix, and the increased likelihood of depression, keeping people well at home can become a real challenge.[37]

Once more, AI and robotics are valuable tools (but not replacements) for healthcare workers. Even something as simple as a chatbot to remind patients to take their medicine or provide some link to their care providers is a potential saver of man-hours and lives. Smart homes also help boost autonomy and freedom by giving them some choice over their surroundings—imagine the frustration of not being able to get up, walk across the room, and turn the lights on or off!

Telemedicine is also a reality, with remote consultation of doctors already a part of daily living. Imagine biosensors that alert you when something is wrong, allowing you to book appointments immediately without needing to travel. Simply discuss it online with a doctor who can look up your vital signs and medical

---

[37] Ibid.

history immediately, and before you know it, the medicine you need is dropped off at your doorstep. Where seniors must live in assisted-living facilities, some are deploying an assistance robot known as Ohmni, which allows family members or caregivers to accompany them remotely, call in medical assistance if needed, and check on their loved ones. All this adds up to more peace of mind for everyone—which is much needed today!

Behind the curtain, AIs are powering research to repair or replace your organs and tissues as they break down, and even improve brain-machine interfaces, one day allowing you to input commands just by thinking about it. Like in science fiction, a machine then responds as if it were part of your body.

Going out is also boosted by the advent of driverless cars. As our reflexes and cognition decline, it makes sense to hand that load off to an AI that takes us to our destinations in comfort and efficiency, and the technology is being worked on as I write. It remains to be perfected, and I know that as today's working population ages, this technology will work better and be more affordable by the time we need it for ourselves. I'm bullish about AI precisely because it caters to our needs as we age, both directly and indirectly through making our younger population more capable.

Provided we live in a sufficiently advanced economy, we're using AI assistance to do more with less so that all of us can optimise our lives and live to the full, whatever our age. Who wouldn't want to live longer and have twenty or thirty more healthy years?

The Fountain of Respect

We're also social creatures, and a vital part of our development is the esteem we're held in by others—in other words, we do care on a deep, profound level what other people think of us, despite what we may say to ourselves. A new generation of online games could help keep us sharp and alert, and a measure of competition included so that we can measure ourselves against players all over the world.

That's where people today grow up getting their dopamine hits, and as we age, we would like our entertainment to adjust accordingly. With screens getting smaller and smaller, what's to stop spectacles or even contact lenses from overlaying an alternate-reality game over the real world? Can you imagine the money a well-planned ARG event would bring in?

We'll discuss the Internet of Things later in this book, but for now, we need to note that all this means voice activation and knowledge of your personal preferences aren't just luxuries but basic features. As more and more vehicles and appliances become voice-activated and easier to use, it'll be a boon to seniors able and willing to move around less. It's simply easier to speak a command than it is to type or even tap it out, and AI is there to learn your voice and preferences, reducing the 'friction' between your wishes and carrying them out. Glen Francis, chief technology officer of Singapore Press Holdings (SPH) says of the result:

> While the media, publishing and entertainment industry has more than its fair share of digital upheaval in the past two decades, even more will come in the next two. The fusion of AI

and immersive technologies will generate an abundance of dynamic and synthetic media content that forms the new mainstream. The lines between live, work, learn and play will be blurred, whilst the physical, digital, and virtual worlds converge.

Of course, all this comes with the privacy concerns we discuss more fully in the Internet of Things chapter. But in a world with Internet-connected devices at home, a clear value transition is taking place. We fear being spied on in our homes less, but adjustment will come slowly and surely.

Remember the writer AI that passed as human, known as GPT-3? As AI and chatbot technology develop, what would a GPT-20 deployed in a care robot or motorised wheelchair look like? Such a programme would be able to predict needs, talk to patients, and comfort them as a human would, but would this then qualify as consciousness?

Of course, we could go into even more disturbing territory and imagine such software installed in sex dolls. In cultures with a history of sex selection, a high preference for sons instead of daughters means men are far more plentiful than women. The laws of supply and demand mean that women can afford to be far more selective about their partners, and there will be far more lonely hearts out there. It would sadden, but not surprise, me to find robots become romantic and even sexual companions. This is not even considering the rise of virtual or augmented reality, which we discussed at the beginning of this chapter.

There's a darker side to this. What's stopping AIs from learning the habits, vocabulary, and mannerisms of celebrities, then enacting them in the form of virtual avatars or sex robots? I would be very surprised if the porn industry did not immediately try to create something like this and even be at the forefront of AI learning and use.

AI-augmented VR experiences will one day be built to engage every sense, not just sight and hearing. Young children are already conditioned to have fun online, and people will move their entire existences into VR when they can. Despite our consciousness being housed biologically, it will move online into new worlds and experiences. Addiction to games and VR will become even more of a reality than it already is, and even substitute for real life. After all, if someone is bullied or is physically weaker in real life, becoming good at an online game is a way for them to command respect they never could in person.

You have complete control over your second life, which is why such games are so popular and will become even more so in the future and be played by people of every age group. Soon, *Memories of the Alhambra* may no longer be science fiction!

FOUR

On the Trail of a Pandemic

> I recommend it to the Charity of all good People to look back, and reflect duly upon the Terrors of the Time; and whoever does so will see, that it is not an ordinary Strength that cou'd support it; it was not like appearing in the Head of an Army, or charging a Body of Horse in the Field; but it was charging Death itself on his pale Horse; to stay indeed was to die, and it could be esteemed nothing less.
>
> Daniel Defoe, *A Journal of the Plague Year* (1722)

There's a saying that information is power, and the COVID-19 pandemic has made that apparent. Urban areas with advanced economies and a lot of cities have a problem that more rural or out-of-the-way ones don't, and that is a higher risk of transmitting the virus. An infectious disease can spread much more easily in a city, where people live, play, and work close to one another.

The climate also plays a part; disease-causing bacteria and viruses, as well as their vector creatures like mosquitoes, fleas, and rats, evolved to thrive in a warmer world. A warming climate means more opportunities for them to spread and run wild for more of the year than they normally would.

Finally, with air travel connecting so much of the globe, the conditions could not have been better for COVID-19 to spread out of China.

One of the factors that led to COVID-19 spreading worldwide was global travel during Chinese New Year in 2020, which, if restricted, could have bought more time for the rest of the world and saved many lives. But while we can learn from the past, we can't change it. The best we can do for everyone is to prepare for the future—nothing stops more virulent, deadlier plagues from emerging. The faster we move, the faster we can isolate the sick and the faster we can return to normal. There's a clear prevention phase, a detection phase, and a resolution phase.

Prevention is where our sensor networks can do the best. Are we measuring the right factors to achieve our goal, and how much uncertainty is there in our readings? Are the sensors able to withstand the conditions we'll place them in?

This is a crucial question to get right because it's the foundation on which we'll build everything else. An error here will cascade down the chain of parts from sensor to software reader to hardware response, and mistakes here have cost hundreds of lives and billions of dollars.

For instance, the most expensive aircraft loss in history occurred in February 2008, when moisture build-up in the humid, rainy conditions of Guam got into an attitude sensor of a US Air Force B-2 Spirit stealth bomber, the *Spirit of Kansas*. Shortly after take-off on a routine flight, the sensor began feeding the bomber's flight computer inaccurate information about its angle of attack and speed, and the computer 'compensated' with an automated nose-up input.

The *Spirit* began to stall, and at that low altitude, there was no time to correct it. Both pilots ejected safely, and the left wingtip struck the ground, causing the *Spirit of Kansas* to tumble over and crash. The fuel ignited, and the subsequent fire destroyed the billion-dollar aircraft.

If prevention can't be carried out, speedy detection when a problem has escalated is the next step—be it a pandemic that has spread beyond its initial zone or an elderly patient getting out of bed and risking a dangerous fall. The faster a response can be summoned, the better.

This means the information we gather must be timely, and even a few minutes' delay can make an enormous difference. It's precisely in this role that AI shines as the amount of data it can sort and interpret in a fraction of a second would take unassisted humans months or years, if they even find those patterns at all.

This allows a novel use in detecting disease outbreaks early. If, say, an Ebola epidemic broke out in a rural area, one of the fastest ways we could find out is if social media in that area began posting about its symptoms all at once, and both ordinary people and doctors began warning about them—even before we have a name to the disease. In *Star Wars: A New*

Hope, Obi-Wan Kenobi can tell when the planet Alderaan has been blown up when he tells us: 'I felt a great disturbance in the Force, as if millions of voices cried out in terror, and were suddenly silenced.' With AI noticing those 'voices', medical authorities can set up speedy quarantines and have resources and personnel made ready to respond faster.

All these advantages, however, *still* rely on human programmers who must 'teach' their software to discern the quality of the data, tell real from fake, and refine their algorithms while still producing the desired result. The AI does the work, and over time, more and more can be left up to it; we merely refine how it does so, managing the process rather than the software itself.

Meanwhile, who watches *them*? When aircraft manufacturer Boeing included a narrow AI called the Maneuvering Characteristics Augmentation System (MCAS) in its 737 MAX airliner to mitigate the pitch changes from its new aerodynamics, the intent was for pilots of the previous generation of 737s to transition easily, serving as an aid to familiarisation of the new type.

However, Boeing had not drawn attention to the new system in pilots' manuals of the 737 MAX, effectively keeping knowledge of the MCAS from the people who most needed to know about it. In 2018 and 2019, erroneous sensor readings led to the MCAS dropping the noses of two 737 MAX aircraft, causing both to crash and killing 346 people in all. The US Federal Aviation Authority acted swiftly to ground the entire 737 MAX fleet until the problem could be sorted out—but it does point to the need for regulators to be both well informed and forward-thinking enough to anticipate these issues.

As the saying goes, aviation regulations are written in blood. If AI solutions are to be of help in detecting and preventing future outbreaks accurately, their development must be carefully watched and vetted by authorities capable of knowing when something has gone wrong, and taking action to correct the problem.

The final phase is resolution, in which we find and implement ways to apply what we've learnt. If we do this right, we continue the upward momentum we've developed over the last thirty years.

AI versus the Plague

Who were the first COVID-19 patients in Singapore? According to the *Straits Times*, it was a tour group from Guangxi, China, that visited for the Chinese New Year of 2020.[38] They saw all the most popular tourist sites and visited a Chinese medicine shop popular with mainlanders like themselves.

The tour went without incident, and the group returned home, but in early February, locals began falling ill. This is believed to be Singapore's first community cluster, and while cases remained low at the beginning, the disease quickly leapt from one patient to another, infecting thousands over the month of April. That led to a nation-wide mobility freeze called a circuit breaker, with daily activities cut to a bare minimum for months.

---

38 Tee Zhuo, 'Coronavirus: China tour group linked to local transmissions visited at least six places in Singapore', *The Straits Times*, 5 February 2020, at https://www.straitstimes.com/singapore/coronavirus-china-tour-group-linked-to-local-transmissions-visited-at-least-six-places-in.

Over this time, one of the world's most sophisticated contact-tracing programmes was implemented, using powerful data analytics to trace chains of transmission from one patient to another. Quarantines and stay-at-home orders were implemented, with anyone who had come into contact with a positive case required to remain in their homes for two weeks.

People were required to scan in and out of public areas, using a smartphone app tied to their identities—or else, an identity card itself. With cases peaking and gradually tapering off, densely populated Singapore has so far avoided the fate of other urbanised countries, and hospitals did not become overcrowded beyond their capacity to manage. There would be an uptick in 2021, but as I write, it is being brought back under control.

AI was a key tool of COVID management from the beginning. Even while it was spreading through China, health authorities were turning to the advantages of AI to curb its transmission:

> Its uses seem to have included support for measures restricting the movement of populations, forecasting the evolution of disease outbreaks and research for the development of a vaccine or treatment. With regard to the latter aspect, AI has been used to speed up genome sequencing, make faster diagnoses, carry out scanner analyses or, more occasionally, handle maintenance and delivery robots.[39]

---

39 Andy Chun, 'In a time of coronavirus, China's investment in AI is paying off in a big way', *South China Morning Post*, 18 March 2020, at https://www.scmp.com/comment/opinion/article/3075553/time-coronavirus-chinas-investment-ai-paying-big-way.

One of the first things to be known about the virus itself was its full structure, predicted by AIs and saving months of research and study. By predicting how its proteins folded, it helped its users greatly reduce the amount of time needed to bring a usable vaccine to market; less than a year after the first cases were reported, companies like Pfizer and AstraZeneca were delivering vaccines all over the world. How effective they will be over the long term remains to be seen, but as treatments are rolled yet in the years to come, their creators will do so with more knowledge than ever, and as I write the traditional regime of vaccines and boosters after a year or so may give way to more easily taken oral pills and other, less invasive treatments.

'Healthcare and pandemic management involve not only the biological, but the social, cultural, and the governmental,' Alan Lee, head of innovation and operations excellence at ride-sharing firm Grab, rightly observes. 'Wouldn't it be great if AI were to be able to advise us like the Oracles of yore, on the wise courses of actions that nations need to undertake?'

What fields were AI driving forward that enabled an achievement like this to happen? The current system of pandemic control is a convergence of several AI-empowered technologies, including:

1. *Computer modelling of the pandemic.*

    A Canadian start-up called BlueDot had an AI examine travel patterns, population distribution, the climate, and much more. When assigned to examine the initial outbreak in Wuhan, China, it identified the virus's path and predicted the cities that would most likely experience serious outbreaks.

Armed with the AI outputs, it's possible that movement restrictions ended up saving even more lives by being put in place while they were of use. Without a chain of transmission, the virus's impact on many areas was less severe than it could have been. With better refinement, a pre-emptive closure of certain state and provincial borders could be carried out to stop a future pandemic in its tracks, before it spreads around the world.

2. *Sharing of knowledge among healthcare authorities.*

Researchers around the world were better able to collaborate on identifying the structure of the virus, its biological make-up, and work on a vaccine. Thousands of research papers were published, and rather than have researchers wade through the entire body of literature, AIs collated the documents available so that clinicians could get quick and easy answers to their questions.

In March 2020, the American government announced a new collaboration with researchers—an AI would:

> mine through the avalanche of research to answer questions that could help medical and public health experts. By cross-referencing papers and searching for patterns, AI algorithms might help discover new possible treatments or factors that make the virus worse for some patients.[40]

---

40 Will Knight, 'Researchers Will Deploy AI to Better Understand Coronavirus', *Wired*, 17 March 2020, at https://www.wired.com/story/researchers-deploy-ai-better-understand-coronavirus.

All this opened up new avenues that would have remained closed had researchers worldwide not been able to pool together all the available information. While current treatments are still not perfect, they are still far better-developed than if research teams had been left to themselves.

3. *Testing and diagnosis.*

AI scanners were trained to detect the signature lung issues caused by the virus, greatly speeding up the screening of thousands of cases. According to Alibaba's research arm, an AI system was trained:

> to recognise coronaviruses with an accuracy claimed to be 96%. According to the company, the system could process the 300 to 400 scans needed to diagnose a coronavirus in 20 to 30 seconds, whereas the same operation would usually take an experienced doctor 10 to 15 minutes.[41]

With COVID-19's rapid transmission, testing of thousands of possible cases became a necessity very quickly—but thanks to AI assistance, the industry was able to keep pace with infections.

As I write, Singapore has greenlit a third national supercomputer, following on from Aspire 1 (completed in 2016) and Aspire 2a, which is scheduled to come online early in 2022. They

---

41 'AI and control of Covid-19 coronavirus', *Council of Europe*, at https://www.coe.int/en/web/artificial-intelligence/ai-and-control-of-covid-19-coronavirus.

are already being put to use in a variety of uses, but the third will be cited within the National University Health System (NUHS) and is expected to focus on healthcare applications. 'The new supercomputer is so fast it is expected to train artificial intelligence (AI) to predict a patient's future disease condition—such as Covid-19 and kidney disease—within hours, instead of days as with standard computers,' notes Kenny Chee in *the Straits Times*.[42]

Uses for it include the development of new drugs, especially to combat COVID-19 and other diseases, study genetics and protein folding in greater depth, and even train AIs to detect diseases, identify obstacles in a robot's path, or even the risk of hospital readmissions—provided, of course, the right data is sorted and collected.

AI remains a relatively new tool in researchers' arsenals, but the questions it raises about privacy, freedom of information, and the impact on individual liberty are still open. With the COVID-19 pandemic, society itself is more divided than ever, with governments repressing their own people's freedoms and other medical services to fight it, using ever more intrusive methods. Sometimes, I wonder if we have more to fear from the uncertainty and lockdowns than the virus itself!

---

42 Kenny Chee, 'Singapore to build third national supercomputer for Covid-19, other healthcare research', *The Straits Times*, 3 December 2021, at https://www.straitstimes.com/tech/tech-news/spore-to-build-third-national-supercomputer-for-covid-19-other-healthcare-research.

FIVE

AI Comes Home: The Internet of Things

> A toaster is just a death ray with a smaller power supply! As soon as I figure out how to tap into the main reactors, I will burn the world!
>
> The Toaster,
> *Fallout New Vegas: Old World Blues*

Imagine you're having a long day at work, and more than anything else, all you want is to come home and relax, but there's housework to do, children to care for, and a quick dinner to make. There's just so much to do between now and when it's finally time to sink into bed, and you don't even think you're rested by the time you have to wake up the next morning to return to work.

Now imagine the same day at work, only you're excited at how much you'll be able to relax. As you're working on your career and your children are being watched over at school, your home itself has been at work for you. Your dishwasher has just been freshly stocked with soap it ordered for itself off the Internet—all you've had to do was take the bottle off your doorstep, open the cap and pour it in, and for the next week every plate, bowl and piece of cutlery will be sparkling clean every morning.

Your clothes sit neatly folded in a dryer, where your washing machine has deposited them. Your trash has been vacuumed up by a next-generation cleaning robot, which has also mopped and sterilised the floor as it navigates around your house, nimbly avoiding tables, chairs, your children's precious toys, and the family cat.

Once your autonomous car is within range, your home prepares to welcome you. (There isn't even a need to enter by a carpark gantry; a unique ID has already been sent to the security system in your apartment complex, recording when that vehicle entered and left.) The lights dim, and soft jazz music begins to play; you've instructed it to surprise you with the latest new recommendations from Spotify. As your face, fingerprint, and retinal scan unlock the door, you nod to the webcam 'eyes' of your home, almost as if such a faithful servant and security guard deserves acknowledgement of some sort. After all, it intimately knows your desires and preferences, and those of your family. If anyone other than your family and close friends show up, it can read the situation and contact you or, if needed, the police.

The home AI actually understands the gesture and emits a cheerful beep in response. It's not a spoken hello—you don't really prefer it, even though such a vocalisation is well within its ability. A report of all the housework it's had the interconnected appliances of your home carry out has been sent to your email, but a quick glance through it tells you everything is in order.

Your family will be home soon, but for now, you have the house—and a bit of precious time—to yourself. Your home assistant isn't the most expensive on the market, but it's got all the household jobs done at a fraction of the cost a human helper would have set you back, but how much is the time to relax and unwind worth?

You walk into the bathroom, take off the smart glasses that have served you so well at work, and look at your face in the mirror, but this mirror can look back at you. Over months and months of analysis and data checks with other mirrors that have seen millions of other faces around the world, it can get a deeper picture of your health and stress levels than ever. Just the movement and shade of your eyes reveal much, but you've elected for it not to stress you out by telling you upfront.

After a quick wash, you settle into a recliner that immediately recognises you and begins to massage your tired shoulders. When your family returns, you'll be there to welcome them back, fresh and cleaned up instead of harried and exhausted from all the housework.

'As the Internet extends its reach into physical objects and becomes the Internet of Things, it will rewire every object in its path. What's considered a futuristic product today will soon become commonplace,' predicts Bruce Sinclair in his book *IoT Inc*.[43]

---

43 Bruce Sinclair, *IoT Inc.: How Your Company Can Use the Internet of Things to Win in the Outcome Economy* (New York, New York: McGraw Hill, 2005), xix.

In this case, the Internet hasn't just given you information you could use to make better decisions. It's actively made your life better by removing much of the stress and repetitive nature of housework, allowing you to relax and present a better side of yourself to your family. Even if you've had a bad day at work, your home is still helping to mitigate the stress, keeping the edge off, and making it easier to keep a cool head. It's actively helping you make life with your loved ones better and, by extension, helping them give the challenges of life their best. What an effectiveness multiplier this technology could be!

Such a scenario is well within reach. 'Years ago, if you told someone that you could talk to your home and get it to control the lights, temperature, locks, or even boil water for you, people would think that you were nuts,' notes Singapore property portal PropertyGuru.[44] Indeed, the IoT is simply a bunch of devices connected over the Internet, 'from simple sensors to smartphones and wearable', according to *techUK* IoT programme head Matthew Evans.[45] From here, it is a simple step to combining this with automation that takes this information, applies analysis and rules to it, and learns through it all—so that users receive better assistance with a particular task as the software observes them and learns their behaviour and preferences.

---

44 '5 Smart Home Projects in Singapore: Why Having a Smart Home Can Make You Feel like a Million Bucks', *PropertyGuru*, 21 May 2021, at https://www.propertyguru.com.sg/property-guides/why-a-smart-home-can-make-you-feel-like-a-million-bucks-and-5-smart-homes-projects-that-will-help-you-16904.
45 Quoted in Matt Burgess, 'What is the Internet of Things? WIRED explains', *Wired*, 16 February 2018, at https://www.wired.co.uk/article/internet-of-things-what-is-explained-iot.

Home AIs have been around for many years and are sold commercially by companies around the world, such as Singapore's own Habitap. In 2018, it and developer Keppel Land unveiled the first AI-powered smart home, which 'currently integrates three core functions of smart home controls, community management and lifestyle services seamlessly on a single platform'. With more features, it will be able to 'progressively anticipate users' preferences and usage patterns, thereby automating features and settings to provide seamless and intuitive experiences'.[46]

Smart Homes for a Smart Nation

Today, smart homes are a multibillion-dollar industry, with an estimated size of US$53.45 billion by 2022. This is more than double that of its worth in 2016, and it's a key investment area for Singapore's vision of a smart nation. Because more than half of Singaporeans live in public housing, the country's Housing Development Board is working with manufacturers to test smart home ideas by integrating them in new apartments.[47]

The concept is already being tested with smart electrical and water meters, which can automatically transmit usage and carbon emissions data to a homeowner's smartphone, as well as smart carparks that do away with gantries and charge cards. Instead, they allow users to freely drive in and out, record their

---

46 'Singapore's first smart home powered by artificial intelligence', *iProperty*, 14 November 2018, at https://www.iproperty.com.sg/news/singapores-first-smart-home-powered-by-artificial-intelligence.
47 Michelle Ng, '1,190 families get keys to first smart-enabled HDB homes here', *The Straits Times*, 31 December 2020. at https://www.straitstimes.com/singapore/housing/1190-families-get-keys-to-first-smart-enabled-hdb-homes-here.

license plates, and charge for parking time to a linked credit or debit card. No offence, but that does make their buyers 'guinea pigs' of a sort, though as we've seen, most people won't mind if the convenience outweighs the cost.

It's one thing to be able to build a smart, responsive home but quite another to pool different services together and make the end result a pleasant, user-friendly experience the customer wants. But if integration with already-popular messaging apps like WhatsApp or WeChat grows, it'll solve that problem by making a home AI just like a friendly domestic helper we can talk to. If homes like this become more widespread, there's no reason for them not to grow more popular in the future and perhaps one day remove the need for human helpers in most cases. Like drivers, their days are numbered—but until the day AIs can automate all housework (which won't happen any time soon), the role will merely change, not disappear entirely.

Off-the-shelf solutions like Google Assistant or Amazon's Alexa are also popular, and machine learning is making such IoT technology more and more convenient every day. If it's electrically powered, chances are some version of it is already part of an IoT ecosystem that can be controlled with just a tap on our smartphones. I've called AI 'fairy dust' because it has the power to revolutionise our daily lives, very quickly—almost as if we were watching Cinderella's coach taking shape before our very eyes. Our cars already rely on small but powerful sensors, but a company that can bring them into the home and balance the convenience with the cybersecurity and programming expertise needed would be very successful. Thanks to competition, the devices themselves are getting cheaper and more widespread, to the point that a non-IoT appliance will end up costing more.

Integration into your home network may very well be just another feature.

Think of the possibilities. One day, there won't be any need to be invasively swabbed to ensure you don't have an infectious disease because sensors in your home and even embedded in your clothes will have sampled any pathogens from the very air. An Internet-connected air purifier might alert you to any pollution in your home, while a smart fish tank automatically measures the pH, temperature, oxygen content, and other properties of the water, adjusting them to compensate for any changes without you having to consistently check them.

A smart doorbell might recognise you and your family even before you arrive at the doorstep and let you remotely monitor your house's entrance; a smart lock might let you into the house at a voice command or via a smartphone app, without the hassle of bringing even an electronic key card with you. Smart homes are effectively basic living sprinkled with AI, and if the adoption of 5G and Wi-Fi 6 are any indication, an explosion of smart IoT devices is completely inevitable.

Anything that can be sensed and connected to the Internet can and will be connected to this vast IoT web, for good or ill. But with this convenience, we pay the price in privacy—we're effectively giving a distant, faceless company access to the most intimate details of our lives, and even the workings of our bodies. If you put on a pair of VR glasses, for instance, even the size of your pupils and the position of your glance are being recorded. Such information could be sold to advertisers who want to know how certain ads and stimuli affect customers.

They'll know from any production when you're alert, when you're looking away, and other subtle cues a human observer may forget, but an AI will not, provided it's looking for the right things.

Abuse and Misuse

It's not hard to imagine how such technology could be misused—writers have been doing it for generations. In his 1949 novel *Nineteen Eighty-Four*, George Orwell imagines a surveillance state that inserts a device known as a 'telescreen' in every home—it is a combination of TV, camera, and microphone, designed to monitor citizens and party members for any hint of rebellion.

Real life caught up when in 2015, concerns emerged over users' ability to give voice commands to the new Samsung Smart TV, raising concerns that private conversations and information like passwords could be picked up by third parties. Like Orwell's hero Winston Smith, who could be spied on at any time and had to watch what he said and did even in his own home, users feared that their own appliances had become surveillance devices.

While the company tried to assure customers that all data it collected was securely encrypted and the voice command function could be deactivated, it failed to address the root issue—that private conversations were being overheard and recorded in the first place.[48] And turning off the TV's voice

---

48 Nicholas Garcia, 'Samsung Smart TV Can Hear What You Say And Record It', *Lifehack*, (no date), at https://www.lifehack.org/articles/technology/samsung-smart-can-hear-what-you-say-and-record.html.

feature and its Internet connection weren't real solutions either because those would have killed the entire point of a smart device.

Anything connected to the Internet can be hacked or misused, and appliances are no different. Information about you may be useful to the company gathering it, but what stops bad actors from getting hold of them or, worse, deactivating your 'smart' home altogether? What if blackouts rear their ugly heads, cutting your smart home off from power? What if a smart lock was defeated by a hacker, or if one took over these crucial functions and turned them against you? Cybersecurity strategist Tom Kellermann warns at CNBC:

> In their current state, irresponsibly implementing these technologies around your home can put your personal privacy and safety at risk. These devices can open the door for hackers looking to spy on you, steal from you, or even actively torment you, as was the case with a Milwaukee couple who came home to find their thermostat turned to 90 degrees and a disembodied voice talking to them through the speakers.[49]

Real people are affected, even if those responsible tell themselves it's being done for the greater good. Here in sunny and often sweltering Singapore, our air conditioners are rarely set to twenty-six degrees Celsius as it gets uncomfortably warm in the enclosed spaces that we often work in. But in parts of Texas

---

49 Tom Kellermann, 'If your home is getting smarter, don't leave it vulnerable to hackers: Cyber strategist', *CNBC*, 20 December 2019, at https://www.cnbc.com/2019/11/30/how-to-defend-your-smart-home-from-hackers-after-black-friday-buys.html.

during the heat wave of June 2021, owners of smart thermostats found that their power companies had adjusted their home temperatures to that setting without their knowledge.

In an energy-saving programme run by the Electric Reliability Council of Texas, 'in exchange for entry into a sweepstakes, customers can allow their utility company to control their temperature during periods of high demand', writes Bill Wadell at *Yahoo News*.[50]

If their thermostats were entered into the programme, the power company would increase the temperature by up to four degrees Fahrenheit to relieve the stress placed on the power grid. While the programme was supposedly voluntary, it turned out many who were on it had not realised it and woken up covered in sweat as their thermostats were adjusted without them noticing.

'One family in Deer Park, Texas, found that their thermostat had manually changed while members of the family, including a 3-month-old girl, took a midday nap,' Wadell says, noting that the girl's parents had raised the possibility of her dehydration. The ability to remotely change thermostat settings appears to affect more than 600,000 of the devices, across twenty-three different states—and this was just by one tech company, EnergyHub.

But with heat waves coming, does it make sense to entrust your home's heating to a distant power company with no way to directly experience the problems that you face? A smart thermostat is meant to give you more control over your

---

50   Bill Wadell, 'Are power companies taking over your smart thermostat?' *Yahoo News*, 26 June 2021, at https://news.yahoo.com/power-companies-taking-over-smart-185552524.html.

environment, but the same people who give it to you also have the power to take it away. If that ability is taken from them by malicious actors, imagine the powers they would wield over customers who believed themselves empowered to control their surroundings. Vigilance by end users is critical, as are updates and constant monitoring by the companies that provide them. As a result, Kellermann suggests keeping them *out* of places that must be kept the safest and most private, such as a child's bedroom or your bathroom.

There is also the possibility of an incident on the company's end 'bricking' your devices, as an outage at Amazon Web Services (AWS) on 7 December 2021 showed. Across the eastern United States, data sharing between the millions of devices that depend on it was suddenly shut down for several hours. Annie Palmer and Jordan Novet of *CNBC Tech* reported:

> Popular websites and heavily used services were knocked offline, including Disney+, Netflix and Ticketmaster. Roomba vacuums, Amazon's Ring security cameras and other internet-connected devices like smart cat litter boxes and app-connected ceiling fans were also taken down by the outage.

Amazon's own internal operations were also affected as they too relied on apps powered by AWS to keep functioning. Palmer and Novet continue: 'For most of Tuesday employees were unable to scan packages or access delivery routes. Third-party sellers also couldn't access a site used to manage customer orders.'[51]

AWS could not even reliably update customers of what was happening, and because the underlying systems were down, affected customers could not even create case tickets asking for help. While its engineers were able to resolve the issue and bring affected services back online, the cost to its reputation and the business efforts of its clients remains to be seen. It's a 'domino' effect, where a relatively small error on the service side has far more serious consequences down the chain.

Stronger Angel, or Fiercer Devil?

C. S. Lewis, the author of the famed *Chronicles of Narnia*, warned that the better and more capable something or someone is of doing good, the worse and more destructive they are when they do evil: 'Brass is mistaken for gold more easily than clay is. And if it finally refuses conversion its corruption will be worse than the corruption of what ye call the lower passions. It is a stronger angel, and therefore, when it falls, a fiercer devil.'[52]

There's always a tug of war between convenience and privacy, and between freedom and security; theoretically, the more information you allow IoT devices and their manufacturers to gather, the more benefits you receive, but the less of yourself

---

51   Annie Palmer and Jordan Novet, 'Amazon Web Services explains outage and will make it easier to track future ones', CNBC Tech, 10 December 2021, at https://www.cnbc.com/2021/12/10/aws-explains-outage-and-will-make-it-easier-to-track-future-ones.html.
52   C. S. Lewis, *The Great Divorce*, in *The Complete C S Lewis Signature Classics*, ebook edition (New York, New York: HarperCollins, 2012), Chapter 11.

you can keep private or separated in case something goes wrong. Governments have a role to play in regulating the transfer of information, but as any history student will tell you, they themselves have a vested interest in learning as much as they can about people and their habits.

This will be a debate for generations to come, but the bottom line is very simple—the digital world, especially the Internet of Things, must augment our abilities and assist us in doing things better. Our children, and their children, will very well have the ability to learn more than we ever dreamed at their age, but that means it is on us to educate them how to do it.

'With AI, network connected devices, nanomachines and other technologies, we can envision the creative destruction of how things are done today. We're equipped with amazingly powerful new capabilities, but all these new technologies need to be managed and controlled,' Alan Lee of Grab points out. 'As usual, the learning journey to adopt that power starts not with the machine, but with the humans.'

With all this convenience, there'll always be room for abuse, and when it strikes, the IoT will bring it closer to home than ever. Let us hope the industry is up to the task.

SIX

The Weapons of the Future

> I struck first. We did all the calculations on how many cruise missiles their ships could handle, so we simply launched more than that.
>
> Lieutenant General Paul Van Riper, USMC (Ret.)

Military exercises are held to keep units ready, and their leaders trained to expect what the next war will bring. But in July 2002, as the American military was in the middle of one war and planning to launch a second, its highest leaders and officials got a rude shock how vulnerable it truly was.

To its credit, the US military knew it was fighting a different kind of war. But the old ways don't leave so easily as the West Coast exercise known as the Millennium Challenge would reveal. 'MC '02 was intended to be the largest, most expensive, and most elaborate concept-development exercise in U.S. military history,' writes Micah Zenko, author of *Red Team: How to Succeed by Thinking Like the Enemy.*[53] It was planned over two years and had grown to involve over 13,000 soldiers, sailors, and airmen—including an entire Navy carrier battle group.

Notably, it was planned to combine computer simulations and live-fire exercises, all of which were aimed at demonstrating what officials called leap-ahead technologies and testing them over a three-week span. Then-secretary of defense Donald Rumsfeld would give it his personal stamp of approval, visiting its headquarters and overall commander, Army General William 'Buck' Kernan, in person. The setup was simple, designed to reflect the then-planned invasion of Iraq the next year. US forces (codenamed 'Blue') would simulate an invasion of a hostile foreign power ('Red'), roughly on par with the potential threats of Iran or Iraq, projected five years into the future.

The organisers firmly anticipated a decisive Blue victory but did not want it to be too easy. The opposing force (OPFOR) commander would have to represent the Red nation at its wiliest and most resourceful, and to that end, Kernan selected retired Marine Lieutenant General Paul Van Riper to lead it. A 'maverick with a reputation for unorthodox thinking', the crafty Marine would make a formidable foe to test new evolutions in American warfighting against.[54]

---

53 Micah Zenko, 'Millennium Challenge: The Real Story of a Corrupted Military Exercise and its Legacy', *War on the Rocks*, 5 November 2015, at https://warontherocks.com/2015/11/millennium-challenge-the-real-story-of-a-corrupted-military-exercise-and-its-legacy.
54 Kyle Mizokami, 'The U.S. Lost a (Fictional) War With Iran 18 Years Ago', *Popular Mechanics*, 3 January 2020, at https://www.popularmechanics.com/military/a30392654/millennium-challenge-qassem-soleimani.

Leading the Blue forces was Army Lieutenant General Burwell B. Bell, and while Van Riper was to do whatever he could within the constraints of the exercise, it was expected that Bell would secure victory in the end. Because of the American technological advantage over almost any other armed force, Blue was allowed all the latest new communications and networking technologies, including some that would not even be deployed till past 2007—and that continue to be on the drawing board, almost two decades later.

Blue began the exercise by demanding Red's surrender. But Van Riper had no intention of sitting back and absorbing the Blue assault and siege of his country—he was going to borrow President George W. Bush's doctrine of pre-emption and strike first. He let Bell's forces know his refusal and laid plans for their arrival. When his electronic communications were jammed, Van Riper transmitted his orders via motorcycle couriers, secret message drops, and even prayers from the 'local mosque'. Then he let his subordinate commanders make their own plans and waited for the Blue naval force to come to him. 'If there was going to be a fight, I was going to get in the first blow,' he would say.[55]

Once the US Navy's ships were fully committed and within range, Van Riper sprang his trap. Massive salvos of simulated missiles roared from their launchers, both land-based and hidden aboard unassuming container ships. Aircraft, flying low and in radio silence to stay hidden from radar and Blue comms interception, loosed even more anti-ship missiles at the hapless battlegroup.

---

55 'The Immutable Nature of War', *PBS*, 4 May 2004, at https://www.pbs.org/wgbh/nova/article/immutable-nature-war.

But Van Riper wasn't done. While the missile swarm captured the Blue weapons systems' attention, dozens of small speedboats made suicide charges directly into the ships, overwhelming their Aegis defence systems and inflicting critical damage.

Van Riper's first strike did not just wound the battle group; it devastated it in less than ten minutes, 'sinking' nineteen ships, including the carrier. In a real-life engagement, thousands of valuable men, officers, aircraft, and other assets would have been sent to the bottom of the Persian Gulf. It would have been the worst loss of US manpower and equipment since the Second World War. Rather than continue the scenario, exercise controllers opted to proceed as if the ships had never been struck and declared Van Riper's surprise attack failed.

'My experience has been that those who focus on the technology, the science, tend towards sloganeering. There's very little intellectual content to what they say, and they use slogans in place of this intellectual content,' Van Riper would reflect in an interview two years later. 'It does a great disservice to the American military, the American defense establishment. "Information dominance," "network-centric warfare," "focused logistics"—you could fill a book with all of these slogans.'[56]

Despite the changes technology makes to the battlespace, the fundamental chaotic, horrific nature of war has not changed. Ironically, as technology makes our lives easier, it also becomes much more fragile as the 'chains' connecting us to the creature comforts we take for granted get ever longer.

---

56 'The Immutable Nature of War', *PBS*, 4 May 2004, at https://www.pbs.org/wgbh/nova/article/immutable-nature-war.

I will tell you right now that I'm very optimistic about tech helping people, but I certainly don't want them blind to how it can ruin them. Rather than sugar-coating or painting rosy pictures that no more match reality than Blue's refloated navy, it's far better to be forthright and transparent now, before the genie is let out of the bottle.

An Army of None

In the comedic novel *Good Omens*, Neil Gaiman and the late Sir Terry Pratchett imagine a meeting between the demon Crowley and his hellish superiors Hastur and Ligur. The two elder demons brag about corrupting a clergyman and a politician bit by bit for years, but Crowley has carried out a far more fiendish operation. 'I tied up *every* portable telephone system in Central London for forty-five minutes at lunchtime,' he tells them.

Hastur and Ligur are flabbergasted. How can *that* help corrupt people and send them to hell? But the unconventional Crowley has no answer they will understand since they are psychologically trapped in the past:

> What could he tell them? That twenty thousand people got bloody furious? That you could hear the arteries clanging shut all across the city? And that then they went back and took it out on their secretaries or traffic wardens or whatever, and they took it out on other people? In all kinds of vindictive little ways which, and here was the good bit, they thought up themselves. For the rest of the day. The pass-along effects were incalculable. Thousands and thousands of

souls all got a faint patina of tarnish, and you hardly had to lift a finger.[57]

The information revolution, for all the good and convenience it has brought, has opened the door to identity theft, pornography, viruses, ransomware, and other evils. But ultimately, they are tools used by humans to gratify human desires. Whether the tool is a rock or a firearm, a printing press or a modern computer, or a human mind or an AI, people will find ways to help or harm one another with it.

AI and computer technology have greatly advanced since 2002, and their military use to cripple an entire country cannot be underestimated. The recruiting slogan 'An Army of One' could well be modified to 'An Army of *None*' because it's now possible to wreck a technologically advanced adversary militarily, economically, and socially without firing a shot or placing a single boot on the ground.

Don't believe me? Imagine losing electricity for just a few days, losing Internet access for a week, or having your water supply cut off for the foreseeable future. This can happen very easily, and it has; in 2011, nearly all the country of Armenia lost Internet access for twelve hours, when an old woman scavenging for copper in neighbouring Georgia accidentally damaged a critical fibre-optic cable with her shovel.[58]

---

[57] Neil Gaiman and Terry Pratchett, *Good Omens: The Nice and Accurate Prophecies of Agnes Nutter, Witch*, ebook edition (London: William Morrow, 2006), 'Eleven Years Ago'.
[58] Giorgi Lomsadze, 'A Shovel Cuts Off Armenia's Internet', *The Wall Street Journal*, 8 April 2011, at https://www.wsj.com/articles/SB10001424052748704630004576249013084603344.

The Red team leaders of today have far more striking power than General Van Riper ever led on those fateful few days in 2002. Developed countries have grown so dependent on cyberspace that any first strike will happen there, and if I were a military planner, I would have spies identify potential back doors and weaknesses in key areas of infrastructure, long before any hostilities break out—not just the big things like power and water but also connection to the Internet and ability to warn one another and transfer data. As Sun Tzu wrote in his classic *The Art of War*: 'Let your plans be as dark and impenetrable as night, and when you move, fall like a thunderbolt.'

Of course, our 'Blue' adversaries will have planned for this in advance, but the element of surprise is a powerful one, and each soldier disrupted, pulled into crowd-control duties or otherwise taken out of the fight, even temporarily, makes a large difference. Ordinary people would already have been conditioned to depend on reliable access to power and water, and the more mass panic and infighting for resources can be sown, the better. Crowley had the right idea; the line between civilisation and chaos is thinner than many think.

Swarms of drones, both air- and ground-based, will be released next, programmed to target radar installations, logistical routes, and rear echelon personnel. During the Second World War, one of the most prevalent threats to Allied shipping (and thus, the war effort itself) was 'wolf packs' of Nazi submarines patrolling the Atlantic and sending valuable merchant ships to Davy Jones's locker.⁵⁹ Swarms of cheap drones stalking an interconnected military force and draining the lifeblood of the country would be the modern equivalent of that, taking valuable fighting strength from the frontline. Destroying a single drone or even a swarm of them will only invite reinforcements as they would all be interconnected in a network that deploys its fighting strength where it is most needed and develop group tactics on the fly. This can all be done very cheaply because for all its lethality, a harassment drone is far less expensive than a soldier. If well supported, such drones are the ultimate commandos, able to stay in an area and wreak havoc for days on end.

Heavier assaults by unmanned ground vehicles (UGVs) would strike more fortified areas, taking them over before resistance can be fully organised. Seen the future battle scene in *The Terminator*? Despite its age and exaggeration of AI capabilities, the depiction of fighting robots of many forms is well done, drawing on the strategy of combined arms. Once a beachhead can be established and an organised force of soldiers and armoured support can be landed, the drones will still be available to launch ambushes and cut communications, further destroying the target country's morale.

It's an arms race, with the prize being the capacity to wage war at maximum risk to the enemy's troops and minimum risk to your own. A small country like Singapore has no choice but

---

59 This tactic was replicated in the Pacific by the US Navy's submarine force against Japanese shipping, with devastating effect. As Fleet Admiral William Halsey would observe: 'If I had to give credit to the instruments and machines that won us the war in the Pacific, I would rank them in this order; submarines, first, radar second, planes third, and bulldozers fourth.' Notice the emphasis on technological advances, the need for delivery systems, *and* building the infrastructure needed to deploy them.

to capitalise on the warfighting advances that AI brings, being small and short on valuable manpower. Cybersecurity will be the first front line of any future war, with AIs causing damage on the scale of any weapon of mass destruction.

AI is the next rung on a long ladder of things people have used to kill one another on a massive scale; we're the only species that constantly invents new and better ways to do this. Whether the tool is a rock or a cruise missile, and whether the delivery method is our own hands or an AI-powered drone, it's people that use them for good or ill. Nuclear weapons have only been used twice, both times against the Empire of Japan—and the horror they unleashed led to the world collectively refusing to use them in a first strike. A policy of mutually assured destruction followed, ironically leading to a more peaceful world order, albeit one fraught with nuclear proliferation into the hundreds of thousands of warheads, near misses, and fear of the end of the world.

I shudder to think about war in the future, and I think an unnamed US Army officer said it best in the uncertain years after World War II. Asked how he thought a third world war would be fought, he simply replied that he did not know. But he did know what weapons would be used in the fourth: 'Sure as hell, they'll be using spears!' he quipped.[60]

---

[60] A slightly reworded version is often attributed to Albert Einstein: 'I know not with what weapons World War III will be fought, but World War IV will be fought with sticks and stones.' For more, see Dan Evon, 'Did Albert Einstein Say World War IV Will be Fought "With Sticks and Stones"?' *Snopes*, 16 April 2018, at https://www.snopes.com/fact-check/einstein-world-war-iv-sticks-stones.

The issue is not what tools are available but how we use them. Put another way, our knowledge must be guided by wisdom and empathy for others; the world only progresses as far as we ease the suffering of others and unlock their potential to contribute to the world around them. Everything that goes against human flourishing also cuts us off from benefiting from those around us. Simple survival and protection from nature and one another occupied the vast majority of people for the vast majority of history, but our generation (at least in developed countries with fair, open governance) has more opportunity than ever to build on the foundation of those who came before us.

Few have had their values as thoroughly tested as former US Navy Vice Admiral James Stockdale. Shot down over Vietnam in 1965 as an attack pilot, he was captured by the North Vietnamese and held prisoner for nearly eight years. Speaking to author Jim Collins after the war, he pointed out that over-optimistic prisoners never made it out of captivity. Those who believed they would be released by Christmas, then Easter, then Thanksgiving and Christmas again ended up bitter and disappointed, their spirits crushed again and again. 'And then they died of a broken heart.'

He continued with the paradoxical lesson he is best known for:

> You must never confuse faith that you will prevail in the end—which you can never afford to lose—with the discipline to confront the most brutal facts of your current reality, whatever they might be.[61]

---

61 Quoted in Jim Collins, 'The Stockdale Paradox', *Jim Collins: Concepts*, at https://www.jimcollins.com/concepts/Stockdale-Concept.html.

In short, we must not give in to fear but acknowledge that there are serious issues to work through and compromise on.

We must unlock our technological capabilities to the benefit of all but carefully consider all the pros and cons that come with them.

We must ask both the easy questions, and the hard ones. The stakes in the world we'll leave to our children, (and their children) demand nothing less.

How does one wire their brain to build the right connections, and resist the wrong ones? My mentor Vikas Malkani puts it in terms of the whole person. 'What are we here for, and how do we create our own happiness and success but without the stress and the struggle?' he has asked. 'Because if we create one element but lose the other, that's not really a good bargain.' Material success and the ability to provide for ourselves and our loved ones is a worthwhile goal—but what if we're so stressed out chasing it that it costs us our health or the relationships we were pursuing it for in the first place?

I came to him to learn meditation and mindfulness, which is fitting since resource-scarce Singapore's greatest strength is the minds of its people. Our founding prime minister, Lee Kuan Yew, wanted to grow a country that was not only dynamic and adaptable but ready psychologically and technologically for the future. Like the companies of tomorrow will do, Singapore was expected from the beginning to bring the future into the present. It's not without its risks, and we need time to slow down, pause, and reflect on where we're going, rather than just how fast we're moving towards it.

All this is good, but social conditioning and inputs also play a role. Hollywood gives us plenty of this in movies like *Terminator* or *Minority Report*, and game studios have conditioned generations of gamers through releases like *Deus Ex* or *Cyberpunk 2077*. All these stories imagine a future where technological growth has merged so much with the human form that it's practically taken over not only how we live but also how we exist.

Perhaps it's best to let Van Riper have the last word: 'A culture not willing to think hard and test itself does not augur well for the future.'[62]

---

62 Quoted in Julian Borger, 'Wake-up call', *The Guardian*, 6 September 2002, at https://www.theguardian.com/world/2002/sep/06/usa.iraq.

SEVEN

The Urgency of Cybersecurity

*Die ich rief, die Geister
Werd' ich nun nicht los.*

('The spirits that I summoned
I now cannot dismiss again.')

Johann von Goethe
'The Sorcerer's Apprentice' (1797)

On Thursday, 22 March 2018, the city of Atlanta, Georgia, found itself in cyberattackers' cross hairs. The city's municipal government was crippled in a ransomware attack—a use of malicious software to shut down its victims' networked computers unless a sum of money is paid to unblock it. With a recovery cost of $17 million, the attack was a wake-up call to the importance of cybersecurity.

The ransomware at work in Atlanta is known as SamSam, which has been around since at least 2015. It is part of a whole family of so-called 'offline' ransomware variants that encrypt files and folders on an infected computer, instead of (or perhaps along with) locking up access to the entire system. The encryption key for unlocking them can be acquired by paying off the hackers, but there are no guarantees they will actually deliver. These attacks often go hand in hand with other types of ransom demands, like those made through WannaCry or NotPetya (both offline/file-type malware).

Cyberattacks are now so common recent reports show that hackers attack a computer in the US every thirty-nine seconds![63] Businesses face over 4,000 hacks every single day using ransomware alone. Add this to the work of cybercriminals who leak the online accounts of their victims, and many people fear being targeted or victimised themselves. According to awareness survey findings released by the Cybersecurity Agency of Singapore, 37 per cent of Singapore respondents reported being victims of at least one cybersecurity incident in 2020.

Cybercrime is also on the rise, with 43 per cent of all crimes reported in Singapore being online ones. This number stood at 16,117 in 2020, a rise from 9,349 in 2019.[64] Most were online scams, although ransomware and botnet attacks remained common occurrences. However, most Singaporeans are both

---

63 'Hackers Attack Every 39 Seconds', Security, 10 February 2017, at https://www.securitymagazine.com/articles/87787-hackers-attack-every-39-seconds.
64 Eileen Yu, 'Singapore sees spikes in ransomware, botnet attacks', ZDNet, 8 July 2021, at https://www.zdnet.com/article/singapore-sees-spikes-in-ransomware-botnet-attacks.

aware of the threat *and* still believe that they will not be targeted by cybercriminals.⁶⁵

Name-brand companies are not spared either, and their very fame means they are constant targets of hackers. Their size is also a vulnerability as thousands of employees have to be practising good cyber hygiene at once, and it only takes one slip-up for a massive attack to succeed and cost the company millions of dollars. Jobs will also be on the line as C-level executives are removed or resign, with line employees also leaving due to the cost-cutting measures that these companies must carry out.

Just as the loot in banks tempts criminals who handle cash, Big Data is a source of illicit gain for the most skilled computer hackers. Cybercrime is an enormous threat to any company, and a major investment in technology could be wiped out in a short period of time. Any data placed on a network, be it sensitive or not, is also vulnerable.

In most cases, attacks occur from outside, but there are also cases where the source is internal to the company, such as a careless or disgruntled employee making a mistake or, worse, engaging in wilful sabotage. Therefore, it is essential to understand how to protect oneself and, above all, how to prevent it.

High-profile victims have included:

- Facebook, which had more than 530 million user records posted on a hacking forum, valuable to cybercriminals

---

[65] Kenny Chee, 'More S'poreans hit by cyber attacks; CSA launches awareness campaign', *The Straits Times*, 28 June 2021, at https://www.straitstimes.com/tech/tech-news/more-sporeans-hit-by-cyber-attacks-csa-launches-awareness-campaign.

eager to impersonate legitimate users so as to scam their friends.[66]
- Credit rating agency Equifax, which experienced a data breach in September 2017, affecting a staggering 147 million customers and costing the firm an estimated $439 million to recover.
- The First American Financial Corporation, which in May 2019 had 885 million records exposed in a data breach that included bank account info, social security numbers, wire transactions, and mortgage paperwork.
- The British National Health Service (NHS), which was temporarily brought to its knees in May 2017 by a relatively rudimentary ransomware attack, resulting in cancelled operations and considerable clean-up costs.[67]
- Yahoo, whose series of data breaches in the 2010s affected every one of its 3 billion customer accounts. The direct costs of the hacks ran to around $350 million, although the total damage to its reputation remains to be seen.

Target: You

In a dramatic scene in Peter Jackson's film adaptation of J. R. R. Tolkien's novel *The Two Towers*, the heroes must convince the passive King Théoden to commit his forces against the darkness threatening all Middle-Earth. It turns out that he has been enthralled by the evil wizard Saruman and rendered too

---

66 Aaron Holmes, '533 million Facebook users' phone numbers and personal data have been leaked online', *Business Insider*, 3 April 2021, at https://www.businessinsider.com/stolen-data-of-533-million-facebook-users-leaked-online-2021-4.
67 'The NHS Cyber Attack', Acronis, at https://www.acronis.com/en-sg/articles/nhs-cyber-attack.

weak-minded to act. 'I would not risk open war,' the mind-controlled king protests.

'Open war is upon you whether you would risk it or not,' the ranger Aragorn reminds him. All of us have the same choice to make as Théoden did because it's not just organisations that face danger from cybercriminals. Identity theft is a huge issue, where hackers steal an individual's personal information and sell it for profit. Victims and their families are placed at risk, with blackmail and extortion being common threats. As we've seen, Internet-connected household items are vulnerable, with hackers even being able to taunt their victims over their home camera's speakers.

A remote workforce comes with myriad dangers, with employees relying on their home networks (and sometimes their own devices) to complete tasks. Often, they must resolve technical issues themselves as IT teams can only help remotely. But with working from home becoming the new normal, cybersecurity is now everyone's responsibility more than ever, and once again, cybercriminals only need to succeed once to cause havoc.

Working from home opens up more vulnerabilities as the same operations have to be carried out without the security protections that office systems provide, such as firewalls and blacklisted IP addresses. With most of our tasks being conducted online, there are far more ways a cybercriminal can compromise them—consider cloud storage of documents, emails and attachments, instant message clients, and third-party services. With so much information being shared digitally, there is so much more to attack.

A cybercrime attack might come through:

1. *A virus* that, once executed, locks the user and others in their network out of crucial files. Usually, a ransom is demanded for a key that unlocks them, in the form of cash or cryptocurrency.

    Other viruses interrupt a company or agency's services through denial of service attacks, which disrupt a targeted organisation's computer network or bring it to a halt. These target servers, distribution networks, or data centres connected to other computers by turning infected machines into 'bots' that overwhelm targets through nonsense requests. The resulting slowdowns could cripple the operations of an important organisation, like a hospital or bank.

2. *Social engineering*, in which people's gullibility, weakness, or carelessness are taken advantage of. Examples include phishing attacks, in which criminals impersonate an authoritative source like a government agency or a bank, and deceive people into handing over confidential information, such as access credentials to bank accounts.

    A distracted or naive employee can fall into the trap and provide emails, passwords, and customer information voluntarily, unaware of the damage they are doing.

3. *Data theft*, in which intruders steal confidential information. This could be the result of a security vulnerability or a successful attempt at social engineering. A *data leak* is often the result of carelessness, while the term *data breach* refers to it being stolen through deliberate action against a person, service, or company.

'With democratization of AI putting game changing computing power into everyone's hands, the bad guys unfortunately are often leading the innovation,' comments Alex Lei, senior vice president and general manager for Asia Pacific and Japan at enterprise cybersecurity firm Proofpoint. 'In cyber security we see a lot of attackers now taking advantage of AI to better attack the defenders, who, for a variety of reasons, underestimate the speed of the change and the potential power AI will bring.'

The Cybersecurity Arms Race

Cybersecurity is simply being ahead in an arms race, in a world where it's easier to do evil than good. Millions of dollars' worth of investment can be undone by a careless victim of phishing emails, a mass scam call, or a carelessly placed thumb drive.

Each time technology solves one problem, it creates a new challenge to counter that solution. Every offence meets a new defence or counteroffensive, and the cycle of innovation continues. Much depends on cooperation between the public and private sectors; there needs to be support and collaboration rather than conflict. This arms race is going to be won by the entities that have a multidisciplinary approach and who draw upon expertise from security experts, threat intelligence providers and academics, as well as those with an understanding of business processes.

AIs are valuable tools in pattern spotting and can find odd new circumstances, like people whose lives haven't changed suddenly making large withdrawals. (This is a telltale sign that the account has been compromised by hackers, or the owner has been conned into making that withdrawal. Even if the cause is completely innocuous, it does merit further investigation.) Their role is what they do best—spotting patterns and alerting their human handlers, even they don't yet understand what those patterns mean.

But we don't engage in it alone because empowered new partners can join us in building a better, more connected yet safer world. It will require new ways of thinking, and more than ever, the private and public sectors need to come together.

Collaboration refers to the idea of working together, often in order to accomplish something that cannot be done by an individual or single company alone. 'Artificial Intelligence is already prominent in cybersecurity. A learning machine that can be taught to defend a network can be taught to attack, and a machine that probes for vulnerabilities without need for sleep or rest can only be offset by another machine constantly on vigil, looking for indicators of an attack,' cautions Ben King, chief information security officer at identity and access management firm OKTA. Rather than have human observers directly keeping watch on threats as they emerge, cybersecurity programmes must anticipate, observe for, and counter them simply because a human operator would be too slow.

'The use of AI in cyber defence is necessary but governing that AI—keeping a human in the loop— becomes the concern,' King adds. In a sense, we find ourselves much like the sorcerer in Goethe's poem; we have an apprentice (our AIs) who must carry out important tasks which we can't constantly supervise, with serious consequences should they go wrong.

But unlike the sorcerer, we can't reverse them with a wave of our arms. The only way forward is to 'train' our AI apprentices in the way they should go, understanding their reactions and how they assess risk and reward, and the ethics of how much penetration they should have into the lives of users. 'Not only does this enable the future of work to be more AI enabled and less prone to errors but will also help and enable in decision making and take away mundane tasks and allow us to focus on things that really matter,' says Alpesh Shindi, DBS Bank's vice president of future of work. It must empower people to get more done, not slow them down—while still protecting them from losing their money to cybercriminals.

'The more we digitalize, the higher the risk of damage in reputation, trust, and profits due to cyber-attacks. It's key that cyber security must be incorporated into the discussions and digitalisation projects right from the start, to make cyber security a key pillar of any digital project,' notes David Ng, country manager at security firm Trend Micro Singapore. 'We need to harness these benefits while ensuring that risk is mitigated and to be able to sleep better at night knowing cyber security systems are in place to take care of our digital avatars, digital assets, social connections, and daily transactions.'

Perhaps nowhere is this balancing act more obvious than in the experience of end users. Anyone who has had their Internet browsing experience interrupted to solve a CAPTCHA puzzle can attest to that; security measures intended to protect websites from AI attacks ironically end up taking away from the end user experience and discouraging humans from using them—and driving them into the arms of competitors who have figured out how to make the process smoother and less troublesome for them.

Striking a Balance

A balance must be struck between what's good for cybersecurity versus what keeps users happy; vulnerabilities will always be there, and every addition, edit, or patch is another opportunity for more to show up, and if not well implemented, users will reject it. Good cybersecurity is like stealth technology; in the same way stealth makes your craft harder to detect and detection itself within the capability of fewer and fewer radars, the right cybersecurity solution makes it such that your organisation's penetration is only possible for those with more and more resources to throw at it.

A large company like Google or Facebook could potentially lose billions of dollars in a data breach and needs larger and more comprehensive measures than a smaller company would. This does not mean smaller companies will slip beneath an attacker's notice; they can and have been attacked, with their reputations lost to data breaches. Attackers choose their targets well, weighing the investment in resources with the potential gain. As the saying goes: 'If you can't catch a fish, a prawn will do.'

Collaboration is the only way forward; it means working together to address an arms race. It requires new thinking and public–private partnerships, like Google's Jigsaw partnering with the Counter Extremism Project (CEP) in order to reduce online extremism.

At the same time, regulators must be in line with the times and creating new conditions, not merely reacting to them. Governments are struggling to come up with legislation that isn't obsolete by the time it comes into effect. This doesn't mean they should forfeit the field to others, like Facebook and Google. Rather, it is a sign that a public–private partnership is more important than ever. The days when security rested solely in the hands of one player are over.

EIGHT

Smart Singapore

> Change is the very essence of life. The moment we cease to change, to be able to adapt, to adjust, to respond effectively to new situations, then we have begun to die.
>
> Prime Minster Lee Kuan Yew of Singapore, 1967

In 'The Machine Stops', a cautionary novella by science fiction writer E. M. Forster, humanity has lost the ability to live on the earth's surface. Everyone now lives underground in vast warrens, spending their days in standardised rooms. All needs, material, emotional and spiritual, are met by an all-powerful Machine, which also allows video chat and instant messaging—much like social media does today. Not a bad prediction for a story from 1909!

But things don't last, and the Machine begins stifling people by banning travel to the surface, and people begin worshipping it and losing the knowledge of how it is to be maintained.

Its repair mechanisms begin to fail, but no one asks why the Machine is going wrong. Instead, they simply decide to trust that the Machine has their interests at heart however bad things get and overlook the faults as being part of its unknowable plans.

Gradually, knowledge of how the Machine works and its repair procedures are lost, and the story ends with the Machine failing completely. Because people have become completely dependent on it *and* incapable of repairing or improving it, civilisation comes to a crashing halt.

As we've developed technologically, it's easy to forget that AIs and the technological marvels we enjoy are at the top of a vast pyramid of human achievement that must be actively maintained. Good, stable governance, the reliable delivery of parts and utilities, and the ability to stay well fed and productive are much like the Machine in the story.

While not omnipotent, it does something much better—it frees us to create, dream, and develop ourselves, rather than merely fight for survival. It also enables travel and interaction with one another in unity and cooperation; we aren't in a Darwinian 'survival of the fittest' scenario, or in the brutal *Game of Thrones*-like world of times past.

Parts in Harmony

Each achievement depends on a complex web of chains, enabling electricity, water, sanitation, logistics, security, and more. Perhaps the founding prime minister of modern Singapore,

Lee Kuan Yew, summed it up best when he declared at the 1984 National Day Rally:

> Everything works, whether it's water, electricity, gas, telephone, telexes, it just has to work. If it doesn't work, I want to know why, and if I am not satisfied, and I often was not, the chief goes, and I have to find another chief.

'It just has to work,' he said—and that reliability and clean governance form the spirit that has made Singapore a research and development hub for the world. That's what will position us at the forefront of AI development and simultaneously keep us one of the safest countries to live and work in. I don't claim we Singaporeans have it perfectly, and our leaders have sometimes been overcautious, waiting when they should have acted. But our strength lies in our ability to *absorb* these mistakes, learn from them, and bounce back.

Singapore's role as a city-state puts it ahead of the curve. UN estimates have just over half of the world's population residing in urban areas, and the percentage 'is expected to increase to 68 percent by 2050'.[68] Here's why, as we adopt new innovations in every sector, we can do so with confidence and safety:

1. *A small, dynamic size.*

    As a city-state of just 6 million people, Singapore's government can quickly implement and respond to global change without the gargantuan bureaucracy a

---

68 '68% of the world population projected to live in urban areas by 2050, says UN', *UN.org*, 16 May 2018, at https://www.un.org/development/desa/en/news/population/2018-revision-of-world-urbanization-prospects.html.

large population would need. Policies can be quickly studied, implemented, and iterated, as was done in the COVID-19 pandemic in 2020.

We've successfully combined the strengths of two vastly different approaches—the central planning and speed to act of China, and the innovative, vibrant start-up culture of the United States. At the same time, we're protecting our abilities for the future through an emphasis on cybersecurity—because this is the only way we can innovate and manage the risk of failure. Innovation without cybersecurity is just a Ferrari without brakes. Our political leadership is also capable of understanding and supporting these needs with the right policies and initiatives.

2. *Trust between private sector and government.*

Both sides have to walk hand in hand, innovating on one end and regulating on the other so that end users are kept safe. The spirit is one of partnership and growth together, not the antagonism one sees in other countries with strong labour union presences. There can be no room for cronyism or corruption, and Singapore's political stability and freedom from these things means a level, fair, and well-governed playing field for everyone.

Because there is trust and cooperation rather than conflict, there is an atmosphere of freedom to experiment, with the knowledge that any AI product must pass certain tests before it can be released to the public at large.

3. *A focus on education and doing well in both book knowledge and innovation.*

We have an education system that's robust and comprehensive and allows students to develop their ability to analyse and discover new solutions, rather than just memorise and regurgitate information.

We do have ways to go in our outlook towards making mistakes and failing, but we're making good progress. Our educational institutions actively pursue overseas partnerships, accelerating change and adopting the best ideas. The debates over lockdowns and mask-wearing aside, it's amazing how quickly we contained COVID-19 in the early part of 2021.

This shows no sign of slowing down, and the development of new technologies and the data that they will use is a key focus of Singapore's government, both academically and in the industry.

4. *A willingness to open up to foreign talent.*

We're a small country and can't survive on our own; it's crucial that the best talents from overseas lend their expertise to us, even if just for a time. We may complain that they take away our jobs, but they're needed for us to improve ourselves and drive change for the better.

That's why, even during the pandemic, business can go on. Our combination of stability, infrastructure, trust, and willpower are crucial to keeping people employed—and if they can't be employed, at least helped to tide over difficult times.

It's not enough to throw money at a sector and hope it improves on its own. Speed and efficiency come from the collaboration of everyone, and Singapore's mix of competent, corruption-free governance; reliable infrastructure; and valuing of good education are like a Machine that allows the engine of human innovation and productivity to work to the full.

Working from home has become the default, and we have our reliable infrastructure to thank as well. 'Who would have imagined we could work anywhere and get our work done without getting out from our nests? In today's world, COVID-19 has accelerated this digital transformation, and employees working from home full-time became real,' says Karen Chong, country manager for Singapore at cloud digital workflow management company ServiceNow.

'With today's digital innovation, AI is changing every aspect of our lives be it at home or at the workplace. We can simplify work with AI and automated services to connect process silos and improve efficiencies by delivering new, digital workflows at any scale.' She is optimistic about how organisations can orient themselves and benefit their team members, rather than replace them:

> Employees can eliminate the burden of guessing of where and how to get assistance with a unified way to request help for anything from any department. AI chatbots can ensure employees can request or get access to information in a simplified manner 24/7. By predicting their

needs and delivering service fast enough, they improve productivity for human team members, and free back-end colleagues from responding to employees' routine requests.

That freeing of team members to fill valuable roles is also how we've kept up research and innovation, allowing for continued development of everything from drones to artificial milk to brand-new farming methods.

The government itself is investing billions of dollars in the process, and it backs promising start-ups through bodies like Enterprise Singapore, smoothening the process of finding venture capital, office space, talent, and more. In November 2021, it launched a new National AI Programme in finance:

> A joint initiative by the Monetary Authority of Singapore (MAS) and the National AI Office (NAIO) at the Smart Nation and Digital Government Office (SNDGO). It aims to build deep AI capabilities within Singapore's financial sector to strengthen customer service, risk management, and business competitiveness.[69]

Among the revealed initiatives are NovA! for analysing financial risk; Veritas to help govern companies using it so they meet fairness, ethnics, accountability, and transparency principles; and others to ensure compliance with security initiatives, issue grants, and share knowledge. The result, it is hoped, will be

---

69 'National programme to deepen AI capabilities in financial services', *Monetary Authority of Singapore*, 8 November 2021, at https://www.mas.gov.sg/news/media-releases/2021/national-programme-to-deepen-ai-capabilities-in-financial-services.

increased productivity and jobs for citizens with the expertise to bring these plans into reality.

AI as a strategy doesn't emerge out of nowhere—there must be concerted effort to support private sector initiatives by a government capable of managing them responsibly and supplying the infrastructure needed in a reliable way. Our advances in AI, supercomputing, and other wonders of technology are the results of that Machine working as it should, not the cause.

Put another way, the 'horse' of good, reliable infrastructure and a culture of building and innovation must come before the 'cart' of new technology that makes life even easier for everyone.

Building the Future

The result has been a country uniquely prepared for the future, and our enjoying the fruits of AI-powered research and development is merely a reflection of that. I call this spirit *pragmatic innovation*, one that recognises, maintains, and understands the Machine that works in the background and allows this to happen. It avoids the twin errors of reckless innovation that ends up crossing ethical lines and harming people, and overly restricted innovation that stifles progress. Every part of our R&D capacity, in both the public and private sectors, moves forward together.

With smart technology fast becoming the norm in urban areas, Singapore is looking to take the lead. As one of Asia's leading smart cities, it will be incorporating AI solutions to make life easier and less labour-intensive. There will be a full technology ecosystem, with suburban areas becoming their own self-contained satellite communities. Some policy changes will include more emphasis on renewable energy sources, fewer cars, and exploration of new technology solutions like smart homes.

It seems that every month now, a new AI-assisted trial is taking place here—from autonomous cars to automated drone safety facilities to robot swans patrolling waterways and even the high-tech military drone called the Arrow.[70] As long as the Machine keeps our daily needs easy to meet and sustains our lives and industries, we can keep experimenting with new technologies and laying the groundwork for the companies that will build the future, such as Tesla, Hyundai, and hundreds of contractor small and medium-sized enterprises (SMEs).

That's how the future now looks smarter and brighter than it has in many years. 'The altruistic goal of any innovation is to improve the life of mankind. From the invention of the steam engine to the invention of computers, the goal has always been to improve productivity and efficiency, thereby leading to better incomes and a higher standard of living for almost everyone,' says computing professor Alex Siow of the National University of Singapore. 'Governments implement these smart cities programmes in transportation, health, utilities, education and more.' All these will be underpinned by AI, blockchain and other emerging new technologies, but with these conveniences come the responsibility for technologists to keep them safe for use.

---

70 Aqil Haziq Mahmud, 'Singapore-based aviation contractor aims to build supersonic combat drone in Seletar hangar', *Channel NewsAsia*, 25 March 2021, at https://www.channelnewsasia.com/news/singapore/supersonic-combat-drone-arrow-seletar-kelley-aerospace-14368346.

A successful 'smart city' leverages the connectivity between people and organisations to lead to two other positive outcomes—inclusion in the system and improvement in everyone's standard of living. Miguel Gamiño Jr, executive vice president of Global Cities at Mastercard, pointed out at the Bloomberg Live 'Sooner Than You Think' forum in 2018:

> We're in this era where 'public-private partnership' is more than just a buzzword. Innovating with the community is important… I've found that when technologies were tested, piloted and iterated in response to feedback in a real, non-sterile environment, the development cycles were much faster and more meaningful, and we ended up with much more robust and resilient solutions.[71]

In other words, smart cities are test beds of new technologies, and iterative tests of all types are constantly being made, allowing new products to be developed and refined every day. This is built on another requirement of smart cities—common open platforms, a wide availability of the sensors and actuators that gather and act on the data, and the infrastructure for widespread Internet access, both wired and wireless.

While creating a smart city often involves the development of new and smart infrastructure, it is just as important to use technology to better leverage a city's existing assets. A smart city needs citizens to be part of its smart ecosystem, which is why Singapore's government has made it a priority to look into

---

[71] 'What Makes a Smart City a Success', *GovTech Singapore*, 5 November 2018, at https://www.tech.gov.sg/media/technews/what-makes-a-smart-city-a-success.

incorporating technology in areas that have traditionally been less tech-savvy.

'In the coming years, the next generation digital technologies, particularly AI, Metaverse, Blockchain and IoT, will combine and reshape every industry, redefine the consumer experience and elevate human creativity to a whole new level,' Glen Francis of SPH has written. 'Smart Nation is not merely about implementing hardware and technologies but, just as importantly, nurturing a technologically sophisticated workforce, innovative business ecosystem and hyper creative culture. Singapore knows what it takes and is making the right investments today to make it happen.'

However, digital technology arises from changing mindsets and skill sets alike. Sau Cheong Siang, CEO of SP Digital, notes:

> Digital is a frame of mind and a way of doing things, beyond just applying technology to improve productivity. It's about re-thinking what you're doing and looking at how to bring new value, being more attuned to customer needs and building and using new capabilities to drive those changes.
>
> Change, as the Greek philosopher Heraclitus tells us, is the only constant in life. Driving digital transformation is not just about applying technology to processes or even training people to certain technologies. To have true digital transformation is to change the mindset altogether and enable the inevitable constant change.

Just because a company leads the market today, it doesn't mean it will still do so (or indeed, even be around) come the next generation. In the 1990s, he personally experienced the emergence of Windows and networking, and how it supplanted the previous popular operating system, Novell NetWare:

> Networking computers became the future, and eventually the Internet took over the world, but a specific technology called NetWare became obsolete and a part of history. To come back on topic on digital transformation and smart nation, as we move forward, we must not be focused on specific technologies or trends but be mindful of where we're heading and why.

One question thus raised is whether coding and, later on, software development, should be taught to everyone in school. To Sau, these are means to the end of creating instructions for machines to execute. It cannot be confined to a single course, but most importantly, the goal must be to solve problems:

> The education on coding today varies greatly. Some courses train the mechanics and how to code, focusing on the mechanics and syntax of the programming languages. Other courses do a broader based training on the fundamentals and theory. Yet other courses focus on the technologies themselves.
>
> Ultimately what's important is not on any single one of the above, but the foundation of problem-solving which is the purpose of software. We develop software and use technology to solve

problems and that must be the underlying focus of any sort of training on coding.

We can choose how much our children are exposed to it, but it must be done to some degree. While not everyone will end up making it a career, it is crucial to make it as ubiquitous as general mathematics and science education, not because everyone should be mathematicians or scientists but because these things are needed in daily life. In the same way, because our lives are dependent on software, it is important to understand how it works.

The Bottom Line

A smart nation framework isn't just a nice-sounding buzzword; it's an umbrella under which Singapore has grouped new initiatives for the future, so the country and its people stay relevant for decades to come. However, it will be hard work bringing it to pass, and years before the technology needed scales up enough to be both widespread and cost-effective.

What our leaders have done is give their true engines of innovation—citizens, educators, and investors working together—the best possible shot they can have at doing so. To join this field is a heavy responsibility indeed!

NINE

The Foundation of Hope

> The management of technology begins with the management of our mind.
>
> Vikas Malkani
> The Wisdom Coach

Note: *I owe the material in this chapter to my mentor and friend Vikas Malkani. He has been instrumental in making my life more stable, spiritual, and successful.*

How do we navigate through a world of noise and new technologies that are impacting us, while staying on point and walking a path that we have consciously chosen to create the outcomes we want? This is where more knowledge cannot help us, only wisdom.

It's only natural—the mind is the origin of technology, so the mind must be managed first before our tools. A famous line from the movie *Jurassic Park* captures it well, spoken when the park's dinosaurs begin running amok: 'Your scientists were so preoccupied with whether or not they could, they didn't stop to think if they should.' How then does the clarity we need come?

True digital leaders must lead others online while not losing their bearings in the real world. The first action is in the digital world, and if you're at the cutting edge of new developments, you must indeed be anchored in that field with knowledge of what you can do and the wisdom to know what you should do. That's what's needed to ethically and responsibly create AI systems that help their users.

But the second anchor that keeps them grounded is in their own mindset—an emotional stability and equanimity that allows them to weather the disappointments and disruptions inherent in life. That's what allows them to make the best use of the technological tools at their disposal, without being used by them.

'It's time for humans need to take start taking responsibility for the AI systems we build. Without mindfulness, humans are little different to robots,' Colin Priest of DataRobot says. 'Competitive pressures may drive us to deliver AI as soon as possible, but let us not forget to stop and observe, to consciously choose the values and goals, to consciously decide how our AI systems should behave.'

For this, we must be very clear about the purpose of our lives because, frankly, we have all the tech we need for basic, physical survival. Think of all the features on an ordinary smartphone sold today—our devices greatly empower us, but how many features do we actually use? For most of us, myself included, the answer is probably 10 to 15 per cent.

So we must now ask, 'What's relevant for *me*?' If I just jump into every new piece of tech that comes along just because everyone's doing it, then I'm just following the crowd. I'm actually moving further away from my own authentic self. People who simply *must* have the latest new smartphone aren't getting real satisfaction when they finally snag it because it doesn't do anything to bring them towards their life's purpose. That desire is only going to disrupt them more in time to come.

Don't Fall for FOMO

At heart is the attitude we call the fear of missing out (FOMO). But keeping up with the Joneses or anyone else is nothing more than a fantasy, and I want to make that clear at the beginning. There's no missing out because there's nothing to miss out on. The latest gadgets and technologies are going to be adopted by some people because it's part of the race they're running against *themselves*. They, you, and I have our own unique race, and there's no need to chase after others' goals.

Remember, the sun and the moon don't compete with each other. They both come out at their appointed times and light up the earth in their own unique way.

FOMO is simply our competitive urge to beat others, turned to the wrong ends. When someone lines up for hours outside the Apple Store to get a new iPhone upon its release—when the same machine is going to be obsolete in a few months, or a year at most—it's not the phone they want but bragging rights among their circle of friends. When they say they're keeping up with Apple, they really mean this. That's a mental issue, not a technological one.

We all need to understand that consumerism can be based on two different forces. The first is what we need; the second is what we want. Now the difference is that while needs are limited and can be sated, there is no limit to our desires. If we are led by them, it's a rat race that will never end.

An old propaganda joke from the Soviet Union epitomises this. In response to criticism from the United States and Europe, the commissars of the USSR devise a demonstration to prove the democratic Western nations wrong. One of them goes to a farmer on a state-owned collective, with a TV news crew in tow.

'Comrade, it's your lucky day!' he greets the farmer. 'Today, your luck has changed. Today, you're going to be the richest man in the entire Soviet Union.'

The farmer is terrified, but he doesn't dare to question the commissar. 'How will that be, Comrade?' he manages to ask.

The commissar sweeps out an arm to indicate the entire tract of land they're on. 'All this land you work, Comrade, belongs to the state. Am I right in saying that?'

'Yes, Comrade. You are right.'

'Well, today the state is going to give you all the land you want—it'll be yours, not ours. All you have to do is run in a circle and cover as much ground as you can before the sun goes down. As long as your start and end point are the same, Comrade, we'll hand over everything within the circle.'

The farmer is stunned and can't believe it, but the camera crews are filming everything. 'Really?' he manages to get out.

'We promise, Comrade. Carry on.'

So the farmer takes off and begins running. He goes slowly at first, but his mind urges him to run faster. *What am I going so slowly for? I'd better run fast and get as much land as I can*, he thinks. He runs and runs, ignoring his hunger and thirst. He pushes himself with thoughts of the future, how rich he will be, and the wonderful life his wife and daughter will have. Despite his body urging him to give up, he goes on and on. (Does this pursuit of an all-encompassing goal sound familiar?)

When he's at breaking point, he realises he's been running for hours, and the sun is going down. Now his greed turns to panic because he has to return to his starting point. He tries to break through his exhaustion and speeds up, but before he can complete the circle, his heart gives out, and he collapses, dead.

Needless to say, the poor farmer loses his life and gets no land at all. Watching all this, the commissar turns to the camera and addresses his viewers. 'You see, comrades? You're not ready to own anything. We're doing this for your good, by protecting you from your own greed.'

It may be Soviet propaganda, but there's a deep lesson here—that because there's no end to our wants, they must not lead us. They can do so to the point where we ignore our needs, to our cost. I'm not for one moment saying making sacrifices for something we want is a bad thing, but at some point, we have to take stock and see if we're doing things the right way.

For a technological example, one need only look at how social media giants capitalise on our need for entertainment, community, and communication with one another. Social media is such an important development because it fills the need for connection with our friends and family, whether it's via Facebook, Instagram, LinkedIn, or something else.

We have a deep longing for community, but precisely how to meet that need is where wisdom comes in, so it doesn't swallow up our time and attention.

Focus on Needs, Not Wants

To find the right centre for your attention, it's important to know yourself and your thought patterns. *In a way, we're all prisoners of the life we've created for ourselves*; we rush and rush to do this and that because getting them done is what we want to do. If you truly want to slow down and pause for a while, nothing stops you. There'll always be time, money, and energy for what we really want because we'll find the resources for it.

This is one of the most empowering concepts I've had the pleasure to be taught by Vikas—that *we're* driven by the lifestyle *we* have created. Nothing, save for our own desires, prevents us from slowing down the pace of life, provided we're willing to earn less. It's doable if we can live on less. For many of us, to downsize is the right size.

Exactly how to change our lives needs to come from wisdom— that is, the right application of a right understanding.

Vikas showed me that the mind has three components, and so does wisdom. Wisdom says that there are three important aspects of life to discover. The first is to know yourself and how you function as a person. The second is to understand how the world works, from the laws of the natural world to the psychology of others. The third is to be able to connect the two—what needs to change before you will change? What changes do you need to make in yourself before you can make changes to the world around you?

'But, Tony and Vikas, isn't time to pause and ask those questions a luxury?' we've been asked. 'We're still on the rat race just trying to catch the next break.'

Why, we must ask, is it even called the rat race? Who taught us to think of ourselves as rats running on a rodent exercise wheel?

Perceptions like this boil down to the way we think of ourselves. In many materialistic cultures, we're taught to aspire to the five Cs for ourselves, and the way to do this is supposedly to get a well-paying job and earn more money.[72] People around us think this way and condition us into being like them; in the same way, the Russian physiologist Ivan Pavlov was able to get dogs to salivate at the ringing of a bell. He taught them to associate the ringing of a bell with mealtime, and so they would get excited and salivate when the bell was rung, with or without feeding them.

We like to think we're more sophisticated than Pavlov's dogs, but how we behave is no less governed by our mental programming. If we want to change our behaviour, it has to start with that programming, in the same way a programmer who wants to

---

72 The five Cs are a reference to the marks of material affluence in a Singapore context. One has supposedly 'made it' when they have acquired cash for spending power, a car which in Singapore is an extremely expensive purchase; a credit card; a condominium, again an expensive piece of private property in a country where prices are high; and country club membership.

make a change goes back to the source code. He must debug it, but this is still much easier than debugging ourselves.

There are two possible triggers for change. The first is inspiration, doing something different and thinking ahead in the absence of a problem. The second, far more powerful one is desperation—when we're forced to change by circumstances out of our control. For instance, a heart attack might make one take up a healthier lifestyle and give up smoking; bankruptcy might be a trigger for better financial habits.

Inspiration says, 'I want to do better.' Desperation says, 'I *must* do better, or I won't survive.' Desperation is more painful, but desperation is also usually the way the majority changes because pain drives people—very few change by inspiration, but it does happen. Inspiration is the better and preferred way.

We don't have to stay in the rat race we were once conditioned to live in, but to get out of it, we need a new conditioning.

Get Grounded in Reality

To solve the problem, we first need to tether ourselves to real life. I talk about social media so much because it's an amazing tool for building the community we need and a touchstone of AI–human interaction, and also a major source not only of distraction from our short-term goals but also a danger to us due to the FOMO fantasy it creates.

This is because what people show the world is not the real them—it's a curated image that they want others to believe is real. Think of a couple whose relationship is on the rocks, yet they'll still post pictures of themselves enjoying life and smiling together. A person posting about a lavish lifestyle might really be in serious debt and actually be *worse* off than someone living more modestly and earning less, but without debt to carry. What most people show the world is not the real them but the image of themselves they want the world to see. An inauthentic person will have an inauthentic feed.

This isn't just true of individuals but of families, organisations, and governments. It's become a means of disinformation and not because of any failing on its part. It spreads disinformation because human beings use it that way. Add to this the fact that you can say practically anything you want, consequence-free; the most Facebook or Twitter can do is block your account, and it's easy to create another. In fact, many controversial writers and content posters do precisely this, creating backup accounts in case their primary one is shut down.

The result is that on social media, it's frighteningly easy to let our worst impulses out. It's made trolling, hypocrisy, and inauthenticity incredibly easy and made keyboard experts out of many an ignorant person.

Why? I believe the root of all this is our need for attention from others. But if that need could be brought under our control, with our minds storm-proof, resilient, and stable, such desires can be controlled to the point where we're in complete charge, not our social media accounts. And when strangers accost or insult us online, it'll just be water off a duck's back.

We don't relieve stress—even great pressure from outside, like many top business leaders face—by wishing the situation would change. Instead, we change ourselves to logically face the reality of our problems. That's why we called it the antiviral mind because whenever negativity is coming into you, it does so in the form of negative pressures, be it stress, anxiety or depression, or even jealousy and FOMO. We protect ourselves by turning those 'viruses' into positive things rather than negative ones.

One simple but effective technique is to realise that whenever you're impacted by someone's negative words or attitude, you're actually focusing on others' judgements about you. Chances are these people know nothing about you, so why should you give it any importance?

One way is to shift direction from their words to the good things already in your own life that bless you. Focus on the benefits and joys of life so the good feelings from them outweigh the bad. When I focus on all the good things in my life, my feelings change as well, and because I'm focusing on the positive, my feelings become positive. It's an attitude of gratitude.

As for the other person, he'll just move on and insult someone else because that's in his nature. I just need to let the storm blow and simply protect myself instead of fighting it.

Building a Mind

Keeping the future in focus, my mentor Vikas and I have created a comprehensive model of how the mind works and how we can train it to become a strong internal foundation in a world dominated by external technology. We call it the *Quantum*

Mind Model© (QMM), in the sense that the mind can be divided into different interwoven but distinct parts, or quanta.[73]

In the same way our computer experience comes from hundreds or thousands of software programmes running at the same time, there are three essential components that make up our minds. All three must be in concert if we are to experience alignment and a sense of grounding within us, and work peacefully with technology in our external life without letting it overwhelm us or lead to a life of distraction.

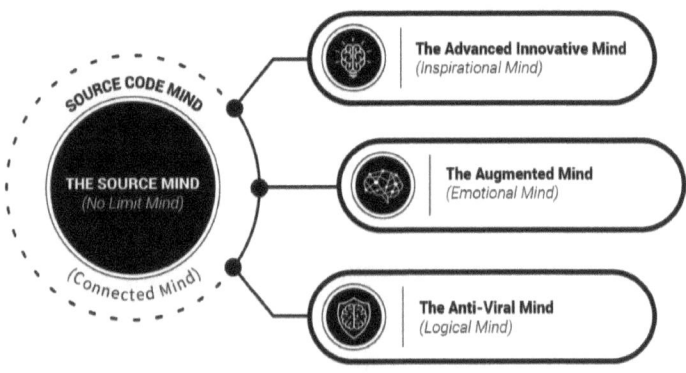

Quantum Mind Model © Vikas Malkani and Tony Tan 2021. All rights reserved.[74]

---

73 Ian Fleming, creator of the famous fictional spy James Bond, also used the term in the sense of human relationships. In his 1959 short story 'Quantum of Solace', Fleming describes the titular concept as the minimal amount of human warmth, empathy, and comfort needed for romantic love to survive. The title (though not the plot) would be borrowed for a 2008 Bond movie starring Daniel Craig.

74 The Quantum Mind Model is a unique model created by and copyright of Vikas Malkani and Tony Tan. All rights reserved. No usage permitted without permission.

In the QMM, we can divide our mind into three parts. The first is the *augmented mind*, so-called because living our daily life augments it with the other two. Because we're born as purely instinctive, emotional creatures, this is the part of the mind that we begin our lives with. It's the primal mind, full of emotions and basic instincts—after all, humans are emotional before they are logical.

The augmented (that is, the emotional) mind is the first part of the mind to operate and develop in us. Think about a child growing up. We're not born capable of logic and higher reasoning, but we feel joy, fear, and more primal emotions. We know when we're hungry, tired, thirsty, or lonely. As we grow up, we have to be taught logic, from 2 + 2 = 4 to language and safely exploring the world around us.

The augmented mind is innately connected to our senses because it's how we perceive the world first. This is where stimulus from the external world comes in, and it always provokes some kind of emotional response first. After all, a child learns largely through its senses and emotions in the first phase of its life. This spans the period from birth up to around seven or eight years of age.

The second aspect of the mind is what we call the *antiviral mind*. This is the logical mind, the second quality that our minds develop as we grow. If our brains are the hardware and our augmented mind is a sensor, our antiviral mind is the software that interprets it based on logical thinking, evidence, and proof. It says with the US state of Missouri, 'You've got to show me.'⁷⁵

In other words, it layers causality and logical connections over our experiences, connecting the dots between cause and consequence. It starts to develop from when we are six or seven years old, up until the end of our lives—and it's what gives us the ability to accept or reject ideas based on the evidence we are shown. When a child learns that behaving well gets rewards, behaving badly gets punished, and behaving dangerously gets them hurt, this is the analytical part of the mind forming.

The third aspect of the mind is the *advanced innovative (AI) mind*. In this section, 'AI' refers to this, not artificial intelligence. This is the third aspect of our mind that develops as we begin to combine the augmented mind and antiviral mind to think about life and its possibilities. It's where our imagination resides and where our dreams, ideals, and visions of the future come from. This is where innovation happens.

When it comes into play, it reminds us to think about the path our lives are taking and adjust it if necessary. When Vikas got the vision to leave his role as a successful CEO and train people in wisdom and mindfulness so they could end the stress, struggle, and suffering in their lives, this was what he was tapping into. When a visionary has an idea and works to develop it, his or her AI mind is what's leading the way. It is what reminds the augmented and antiviral minds of what's at stake so we can put

---

75 The expression is the state's unofficial motto. It comes from a saying popularised in 1899 by then-representative Willard Duncan Vandiver, who said in a speech: 'I come from a state that raises corn and cotton, cockleburs and Democrats, and frothy eloquence neither convinces nor satisfies me. I'm from Missouri, and you have got to show me.'

our emotions and doubts to one side and hustle along. It's what keeps us going when our perceptions of hunger, thirst, and tiredness urge us to stop.

All three parts of the mind have their origin in what we call the source code mind. This is the part of our mind that's connected to the universal mind. We could also call it the cosmic, no-limits, or God mind. Once we are trained to connect with the universal mind, all the three parts of our individual mind are energised and empowered. Innovative ideas flow effortlessly, we use logic and analysis to solve problems, and we are emotionally confident and driven to achieve our goals.

When the three parts of our individual mind work in tandem, breakthroughs happen, discoveries are made, and eureka moments become second nature.

I like to think of it as a diamond, with brilliantly shining facets that you can admire by themselves. But every facet comes from the material itself, and the way each carbon atom is bonded to four others to form an incredibly beautiful jewel. (The graphite we see in pencils is also a product of carbon and thus is atomically identical to diamond, but because each atom is only bonded to three others, it has entirely different properties.)

The source code mind is like the operating system for the miraculous hardware that is our brain. You could also visualise a plant, branching out from a central stem into different branches. Each 'branch' represents a part of the mind, but they are all from the same plant. And the plant is grounded in the earth from which it derives its nourishment and strength.

In software, the source code is what creates everything that the computer running it puts out. It's the instructions that the computer simply compiles, reads, and follows. At root, our source code minds connect to the universe, and when we learn, discover, or create something, we're bringing into reality a connection that already exists.

Even great scientists have repeatedly used the word *God* to refer to the cosmic mind and describe the underlying mechanism and harmony of the universe, even if they don't have the religious idea of a personal deity in mind. This was perhaps what Albert Einstein meant when he told a student named Esther Salaman in 1925, 'I want to know how God created this world. I'm not interested in this or that phenomenon, in the spectrum of this or that element. I want to know His thoughts; the rest are just details.' Note that we are speaking of the part of our mental make-up that is in tune with our deeper selves and the universe, and we're limiting our exploration to this sense. Exactly *what* the God mind comprises is beyond the scope of this book and more appropriately belongs in the realm of religion.

My point is that Einstein knew that nothing he discovered was his own invention—he was formulating what already existed. He was tapping into the source mind (or no limits mind) and using his source code mind (connected mind) to download what already existed, and this then became a new discovery, invention, or solution for the world. Any one of us can do this, even if we'll interpret it differently based on our knowledge and experiences. The more we focus on something, the more realisations we will have about it. It is when that source code mind is connected to source mind and the three parts of our conscious mind are working in tandem that we have a foundation on which our understanding of new technology can be built, without the disruption it can bring.

The easiest way to awaken and align the human mind in all its facets is through the practice of meditation and mindfulness. Wisdom is an essential part of this process, and silence and stillness and powerful force multipliers too. If one studies the lives of famous scientists and inventors like Nikola Tesla, Thomas Edison, Einstein, Srinivasa Ramanujan, Copernicus, and Archimedes, one cannot escape the fact that they made great breakthroughs in phases of deep contemplation and silence, when their focus was turned inward. This is what meditation trains us to do.

The Creative Ability

Electric car manufacturer Tesla is named for the great inventor Nikola Tesla, whose innovations continue to astound us even today—some of his alleged developments have never been replicated, even with modern technology. Like Einstein, he didn't arrive at his discoveries with purely analytical thinking but produced them from a place of silence within himself. Instead, these great innovators would study all they could about a particular subject, including the work of earlier scientists, and only then retreat into a place of silence.

Notice that successful silent contemplation doesn't start from zero. It starts as far ahead as possible, which is why we must study whatever material is out there. The real innovation happens when you stop studying and start contemplating—meditating on the subject and taking mental 'steps' forward from the end of your research. This must happen without distractions from here, there, and everywhere.

You don't need to be physically alone, but you do need to get mentally alone. For example, Tesla would lock himself in a hotel room, only allowing food to be brought in. During those sessions, he never even let cleaners in, saying it would disrupt him. On the other hand, Einstein was almost never by himself, but he could retreat into a private space and lose himself in thought—to the point he couldn't even hear it if he was called.

We call a state like that 'flow', as coined by the Hungarian American psychologist Mihaly Csikszentmihalyi. In his 1990 book *Flow: The Psychology of Optimal Experience*, he describes it as 'a state in which people are so involved in an activity that nothing else seems to matter; the experience is so enjoyable that people will continue to do it even at great cost, for the sheer sake of doing it'. In this state, our prefrontal cortexes are de-emphasised as the brain communicates freely within itself, and curiosity and creativity can come to the forefront. It's such a state of integration within the brain that I actually call it connection. It's where we connect with the true mind behind our minds.

Everyone has different ways of connection. It is said that in ancient Greece, the philosopher and mathematician Archimedes was finishing a day of deep thought about the physics of buoyancy and displacement and why some things floated while others sank. The moment he detached from worldly noise and activity, and lowered himself into his bathtub and saw the water he was displacing, the answers he sought suddenly came to him. 'Eureka!' he shouted. ('I've got it!') According to the legend, he was so excited that he leapt from the tub and ran naked into the street.[76]

Of course, anyone who overheard or saw it would have thought him a madman. But who cares?

Notice that the process includes not only doing the research but also *stopping* it. We'll know it's time when what we call our subconscious mind tells us—because it has a way of emerging when, for instance, we're tired and go to sleep. We see this when we're dreaming, and we accept any weird, out-of-place event, location, person, or thing because our emotional and logical minds have 'shut down'.

What sleep does is take us into the 'source code' of our minds, after the noise of the waking world has quietened down. It's easier to get answers or direction in a silent state of contemplation, when you're connecting not to the world but to its very source. If you're religious, you could call it God.

Never Stop Learning

We do our best work in a state of flow, even ignoring our need for food, sleep, or rest if it goes on too long. This was *meant* to be the way people lived, and as such, we're naturally curious and seek to learn more and more throughout our lives.

What we want is conscious lifelong learning, and that means choosing every day to find new things to learn. The brain

---

[76] A popular joke gives his neighbour's witty response: 'Hey, you don't smell so good yourself.'

naturally adapts to this new learning by growing and developing, till the day you die.

Classical theory once held that in the same way the body grows and reaches its prime in twenty to forty years before it starts declining, the brain also has such a cycle. But scientists have since discovered that while this is partially true, the brain is far more 'elastic' than once thought, able to keep growing with age. It can keep growing new neurons and connections for far longer than we thought, and the process may help stave off neurological conditions like Alzheimer's disease.

The mind is flexible, adaptable, open, and stretchable. The mind can evolve to keep pace with new technology, both understanding it and wisely seeking to draw out its benefits while watching out for the costs. This is the foundation on which any reaction to AI, or indeed any new technology, must be built.

We saw earlier in this book that the challenges will be tremendous. The nature of relationships is changing, as previous generations and their more in-person style of community are replaced by younger ones more adept at long-distance communication and a more distant-yet-close digital one. Many of them can chat over social media messaging, but fewer and fewer can have deep, interpersonal conversations. Body language also becomes more and more of a mystery as we shrink our in-person social circles and interact with others less.

Therein lies the danger of becoming a purely reactive creature, always at the mercy of the environment. People simply become a collection of words and pictures on a screen. You've seen how rude they can be to one another on social media, saying things they never would to the same person face to face.

We forget about empathy and compassion. We forget about body language. We forget about connection and emotion. Our relationships have become more challenging in ways they never were before.

That was even before COVID-19 tore through societies all over the world, not only ending lives but tearing people apart as countries enacted lockdown protocols, mandated mask-wearing, and caused businesses and schools to close. We've not only been stripped of so much, but we've grown more disconnected than ever. Perhaps mask-wearing, where we hide our faces and prevent people from reading our expressions, is a stronger symbol than we ever thought!

Again, everything boils down to the mind. Social media appeals to the augmented mind because it elicits strong emotions in us. If we temper our emotions with a strengthened antiviral mind, we can use it without fear—because we're secure in ourselves, we can afford to be open to others. We have the ability to engage with new people, opportunities, and technologies without letting them overwhelm us or becoming addicted to them.

This is easier said than done, however. What makes this so challenging is the satisfaction we get from others liking and sharing our posts, or commenting favourably. Every time it happens, we get a small hit of a neurotransmitter known as dopamine, and if we don't watch out, it can become an addiction. We begin to tie our happiness and sadness to whether other people liked or didn't like our posts—and even get agitated if good friends somehow don't do so. But none of this is real. It's our neurochemistry playing tricks on us.

Now let's be clear that these dopamine hits are an essential part of our self-esteem, morale, and resilience to get through life. They're effectively what make daily living bearable, but when we get it from social media, we put our happiness and peace at the mercy of others. We depend on approval from them, rather than ourselves; we make ourselves not their friends but their prisoners. We've handed to them the power to make us feel good or bad.

As I've said before, we must never forget that the purpose of technology is to make our lives better. If our use of technology, social media, and 24/7 connectivity are not bringing more happiness, fulfilment, and peace into our lives, something has gone wrong. We need to review and reset. We need to course-correct.

Another subtle danger whose poison works slowly over time is procrastination. Because we need a dopamine rush, we turn to our devices again and again for it, putting off important work for just one more check of our social media accounts. What we're after isn't really our friends' updates but the satisfaction of dopamine release. How much great work has gone undone because of this? We know what we're made to do and what we should do, but procrastination makes us delay taking the action needed—and the days have a way of turning into weeks, the weeks into months, and the months into years.

The management of technology begins with the management of our mind. After all, if we can't manage ourselves, how are we going to manage these powerful tools at our disposal? Without the mental strength to stay in mastery of our tools, we will certainly be mastered by them.

They might *disrupt* us so much that we can't function well, like what smartphones and entertainment on demand have done to our attention spans; they might *depress* us by delivering bad news via social media, augmented by AI that simply does as it's programmed and delivers us more of what we like, or they might *desensitise* us to the needs and value of others. We end up talking and showing off to others more than actually doing the hard work of understanding people and helping solve their problems.

Guarding Your Mind

It's indeed fitting that technology can aid us in this respect, and it's not for nothing that mindfulness and meditation training has become a billion-dollar industry. The pressure to speed up, do more, and be more can be exhausting, and although it can help motivate us to heights we were unaware we could reach, we run the risk of ending up like the unfortunate Soviet farmer we met earlier. The 'commissars' who push us beyond our limits may be in our heads now, but that does not make them any less real.

To their credit, companies big and small know this, and invest in their team members' mental health—a list that includes some of the largest tech firms in the world, like Google, Sony, Facebook, and more.[77]

That said, our managers can only do so much. To get much benefit out of it, we ourselves must actively seek those times of stillness and rest. There's nothing religious about this, and mindfulness is simply the process of bringing your attention to the present moment and acknowledging your thoughts, feelings, and sensations. Mindfulness does not require any particular beliefs or behaviours; it's a method for training attention and awareness.

Mindfulness may be beneficial for wellness because it encourages people to live in the 'here and now', which can reduce stress, improve mental acuity and focus, and lead to a greater sense of connection to others. Mindfulness also helps us develop an attitude of acceptance towards our thoughts, which can decrease anxiety and depression.

Finally, mindfulness may also help you feel more grounded in body. We can use it to consciously develop the right habits, such as eating more healthily or by clearing our heads for a few minutes each day. There is actually evidence that mindfulness training changes our brains for the better, and according to the *Harvard Business Review*:

---

77 Marissa Levin, 'Why Google, Nike, and Apple Love Mindfulness Training, and How You Can Easily Love It Too', *Inc*, 12 June 2017, at https://www.inc.com/marissa-levin/why-google-nike-and-apple-love-mindfulness-training-and-how-you-can-easily-love-.html. See also: '6 Companies Using Meditation for a Productive and Happier Workplace', Journey, (no date), at https://journey.live/6-companies-using-meditation-for-a-productive-and-happier-workplace.

When practiced and applied, mindfulness fundamentally alters the operating system of the mind. Through repeated mindfulness practice, brain activity is redirected from ancient, reactionary parts of the brain, including the limbic system, to the newest, rational part of the brain, the prefrontal cortex.

In this way mindfulness practice decreases activity in the parts of the brain responsible for fight-or-flight and knee-jerk reactions while increasing activity in the part of the brain responsible for what's termed our executive functioning. This part of the brain, and the executive functioning skills it supports, is the control center for our thoughts, words, and actions. It's the center of logical thought and impulse control. Simply put, relying more on our executive functioning puts us firmly in the driver's seat of our minds, and by extension our lives.[78]

By putting the pressures of life on pause and focusing on the present moment, we 'practice' inviting our antiviral and advanced innovation minds into the situation we are in. We become more aware of our surroundings and are able to partially 'detach' and make clear-minded decisions rather than the first, instinctive ones that our augmented minds come up with.

---

78 Rasmus Hougaard, Jacqueline Carter, and Gitte Dybkjaer, 'Spending 10 Minutes a Day on Mindfulness Subtly Changes the Way You React to Everything', *Harvard Business Review*, 18 January 2017, at https://hbr.org/2017/01/spending-10-minutes-a-day-on-mindfulness-subtly-changes-the-way-you-react-to-everything.

Technology can be helpful for promoting wellness by providing guided mindfulness sessions. The official Apple Watch mindfulness app even allows meditation periods as short as one minute so that, if needed, you can sprinkle them throughout the day.[79] Headspace is another great app that offers access to free 'mindfulness moments' and paid plans with additional features. Calm also has an interactive website where users can listen to meditations while following along on their computer or mobile device, and Meditation Oasis provides high-quality meditation files for download to your computer or mobile device.

Another great mindfulness tool is the Muse meditation headband, which measures brain activity and provides feedback on how well you are meditating. The goal of using the Muse is to help you learn how to focus and achieve a calm mind. If you are looking for a more quantifiable way to measure your wellness, this tool is a great option.

Strengthening the Foundations

Today, technology is progressing and advancing faster than ever before, but it has always been so if you look at history. Maybe the speed is faster today, as more and more technologies are brought to bear, but it has always been evolving—and changing us in the process. By changing us, technological innovations have changed entire civilisations.

But there's nothing to be afraid of as long as we manage ourselves first. Even the most intelligent AI is simply a tool; after

---

79   Erica Sloan, 'Apple Watch's New 'Reflect' Feature Is Perfect for One-Minute Meditations On-the-Go', *Well and Good*, 1 July 2021, at https://www.wellandgood.com/apple-watch-mindfulness.

all, in most well-governed societies, we instinctively prefer the company of a police officer with a gun to that of an unarmed lunatic who might lash out violently at anyone. The police officer is capable of doing far more harm more easily than the lunatic, but that ability is under control. The lunatic has far less capability to harm others, but we get away from him because what little there is is random and unrestrained. (There are, of course, bad cops who abuse the power they have; in which case, C. S. Lewis's warning from Chapter 5 applies. The better the good something is capable of, the worse the harm it can also do when turned to the wrong ends.)

What was needed, clearly, was a way of thinking that was comprehensive enough to help us navigate the world of tech in every part of life with inner peace, while simple enough to bring to mind at any time. We had to tap into all three areas of the mind, and thus, the idea of our Quantum Mind Model© was born.

As you engage with the ever-increasing technology around you, remind yourself that the purpose of technology is to build a better life. Ask yourself: what is a better life for me? Ask yourself if all the technology that's in your life is making your life simpler, easier, happier, and more joyful. If your answer is yes, you're in a place of control. But if your answer is no, it's time to reclaim control over the technology that's controlling your life.

Knowing what is enough is also important. For example, if you love dogs, having one is joyful, maybe even two. But how about five? Will having more dogs than you can effectively care for be a source of more joy or more stress?

Training your mind is like building a strong foundation that your life grows upon. What we've shown you is but a sampling of that foundation and how the right use of our mind brings it to life. After all, every building's construction begins with a foundation—a strong one will weather any storm and allow rebuilding afterwards, but a weak one will collapse no matter how well the building is dressed up. You can renovate or remodel the building, but a strong foundation will always endure.

As technology advances and wellness becomes more prevalent in our society, we will see an increase of mindfulness tools available to us. They can eventually help us train our minds by recognising negative thoughts, calming ourselves in the face of anxiety, and increasing awareness about how to take care of ourselves during difficult times. It's in the hard times that we need to centre ourselves more, not less.

CONCLUSION

The Human Mind, Unleashed

> We have an opportunity for everyone in the world to have access to all the world's information. This has never before been possible. Why is ubiquitous information so profound? It's a tremendous equalizer. Information is power.
>
> Eric Schmidt

Robotics and AI have advanced so fast that we can't imagine life without them, compared to just a generation ago. The more sensors we have, the more information we'll get and the more we can do with it. Acting on that wealth of information is what gives it its power. There'll be teething problems and times when the system works when it's not intended to—for instance, it was discovered that when an iPhone or iWatch fell and hit the ground in the wrong way, its software would automatically call emergency services if the user did not cancel it.

Artificial intelligence is now a common part of our lives. The Internet of Things has led to an increase in cybersecurity threats as more devices are connected to the internet. As AI becomes ever more embedded into our everyday lives, we must be aware of how it can change us and what ethical dilemmas may arise from these changes. The discoveries we make could potentially save hundreds or even thousands of lives each year, but it also brings up the question of how much we should know about the future.

If AI can predict outbreaks with high levels of accuracy, is there a chance that this information could be made public before it happens? With so much data at our fingertips thanks to the Internet and advancements such as big data analytics, how far into other people's lives should we delve?

If we had access to more data, doctors and care providers might be able to make better decisions based on this information. But this opens up a risk that too much knowledge about other people's lives could lead us to become over-intrusive, losing their trust and, worse, having people hide their symptoms from healthcare providers. That would defeat the entire purpose of learning more about their health and taking care of them!

The future of artificial intelligence looks set to change us all in ways we cannot predict yet. With so much potential for the technology still left unexplored, there is no way of knowing what changes will be made or how these advancements could affect our lives further down the line.

'AI is powerful beyond imagination; in fact, we are just touching the tip of the iceberg!' cautions Benjamin Pheong, security services manager at IBM GCG. 'However, to harness it's capabilities we need a strong framework of regulations and ethics augmented with mindful deliberations to ensure its usage is steered to the benefit of humanity.'

Convergence and Iteration

There's of course room for improvement in every area of AI. What needs to happen is a *convergence* of technology, where various fields come together in new and exciting ways—with a constant, iterative improvement over the years that both looks to the future and what can happen and looks to the past for lessons to be learned. As the saying goes: 'The optimist invented the airplane; the pessimist invented the parachute.'

This is what will determine the future as we stand on the edge of a technological precipice. Not everyone is wise enough to weigh all the evidence and debate the various paths forward, and not everyone even has the luxury of being able to do this; for more than 2 billion people around the world, especially in sub-Saharan Africa and many other poverty-stricken regions, the struggle for even basic survival remains unchanged.

If we don't solve these problems and reduce the inequality gap, it will only be a case of the rich and advanced benefiting, while the poor get nothing at all. William Gibson, author of the visionary novel *Neuromancer*, put it well: 'The future is already here. It's just not very evenly distributed.'

But the good news also if we harness the latest technology and we are mindful about regulating it and putting it to the right uses, we have a chance to reverse the harm done . . . and build a better future together. With our new technological powers, we want to do on a global scale what the nineteenth-century clergyman John Wesley urged his Methodist churches:

> What way, then, (I ask again) can we take, that our money may not sink us to the nethermost hell? There is one way, and there is no other under heaven. If those who 'gain all they can,' and 'save all they can,' will likewise 'give all they can;' then, the more they gain, the more they will grow in grace, and the more treasure they will lay up in heaven.

Nothing happens by chance. Our skills and talents, and the opportunities we have, are shaped by the choices we deliberately made months or years ago. Call this karma, God's will, or fate, but we earn what we work for. Whatever it is, it's a creation of your own choices. Everything we do gives off energy, and it's up to us whether it's positive or negative.

Your thoughts form your habits, which form your life. That's what makes clarity, emotional control, and wisdom so important.

AI and the Taking of Jobs

Will AI take away our jobs, as many people fear? Perhaps a sobering case study will settle the question, one that involves a near miss with a software we looked at earlier in this book—the MCAS system installed in Boeing's 737 MAX airliners.

On an earlier flight before one of the fatal crashes it contributed to, the aircraft involved experienced an MCAS error in the air and went into a dive. Its panicked crew were assisted by an off-duty pilot who happened to be on board, and the problem was diagnosed and power shut off to the motor that was forcing the nose down. The workaround saved the lives of everyone on board, and the aircraft continued to its destination without incident.

But for some reason, the fault went unreported on the ground. A new crew took over the aircraft to operate its next flight, and the same problem arose. With no one aboard able to correct the issue, the airliner dove straight into the ocean, killing all 189 people on board.

Had MCAS taken anyone's job? Certainly not! It had a specific task to perform, but even more human oversight was still needed—coders and engineers to ensure it was working correctly and integrated properly with the aircraft's other systems, pilots, and aircrew to understand the various situations that could arise, and ground personnel to log faults, address them, and bring them to the right people's attention. The disaster drove home the lesson that while an AI performing its assigned task in a dangerous (to us) manner had caused the initial problem, the crash was ultimately the fault of humans who had failed to manage it correctly.

In short, AI development won't take jobs away but change the way the various tasks that constitute them are carried out. DataRobot's Colin Priest points out the principle of AI tasking being designed around helping humans: 'Rather than replacing our jobs, computers have created new jobs and made existing jobs more human-centric, as we delegate tedious mechanistic tasks to machines.'

His main point is that AIs by definition cannot do jobs—they can only do the task assigned to them. The role that we call a *job* actually comprises multiple tasks, each of which might be handled by one or even several different AIs:

> The future of work will be transformed task-by-task, not job-by-job. By analysing which tasks will be automated or augmented, organizations can determine how each job will be affected. Rather than replacing employees, a successful organization will redesign jobs to be human-centric rather than process-centric.

AIs are tools that do a specific task very well, such that we eventually can outsource the repetitive and undesirable tasks of information storage and data analysis, sorting, and manipulation to them. What they do is free us up to handle what we do best, and that is to work alongside others at solving the problems that matter, using our strengths of communication, engagement, and empathy.[80] We're effectively enabled to be *more* human, not less.

Alan Lee of Grab puts it well: 'If AIs wanted to take over our rule-based, transactional jobs, we haven't got a chance. Be smarter than them!'

> Don't focus on areas within your role that can be performed repeatedly, the same way, repeatedly. Chances are, those are the boring parts you never really enjoyed doing anyway.

---

80 Colin Priest, 'Humans and AI: Why AI Won't Take Your Job', *DataRobot*, 21 September 2021, at https://www.datarobot.com/blog/humans-and-ai-why-ai-wont-take-your-job.

If you've been putting off plans to digitalise, innovate, or adopt AI technologies, think! That 'better day to do this' won't ever come when you're no longer existent or relevant, and the time to act is *now*.

We do best when we are continually adding value and learning more and more throughout our lives. It's in our nature to evolve and adapt to our environment, and the new technological landscape we find ourselves in is no exception. In the end, all this tech is only a tool. Whether we create the value that's needed is up to us—as well as the world we leave behind for future generations to steward. It depends on what we do right now.

As the saying goes, the best time to plant a tree was twenty years ago. The next best time is today.

AFTERWORD

AIs and Conscious Thought

Over the past decades, computers have broadly automated tasks that programmers could describe with clear rules and algorithms. Modern machine learning techniques now allow us to do the same for tasks where describing the precise rules is much harder.

Jeff Bezos

Author's Note: *Portions of this book were written with the assistance of Jasper.ai, a writing tool based on the GPT-3 AI writing framework. The afterword that follows, despite being from a human point of view, is purely Jasper's own writing—developed over much analysis of the Internet and, with it, human writing patterns. My only inputs were keywords, a rough direction, and some light editing for clarity.*

Will AIs ever make the jump into self-awareness and consciousness? We can see how divided people are on the matter. The first person believes that artificial intelligence will never be able to have self-awareness or conscious thought. Humans are far superior in every way, because our brains are able to create consciousness, so much so that intelligent machines will never be able to achieve the same level of thought.

The second person believes that artificial intelligence will be conscious, but it won't necessarily share the human experience. Instead, there are some things that self-aware computers can do better than humans because they don't suffer from common human issues like self-doubt or social anxiety.

Finally, the third person believes that there is no way to know if artificial intelligence will be conscious, and we should simply continue experimenting with computers until we reach a point where we can create machines that are as intelligent as humans. It's possible, and even likely, that one day humans will create a computer that is self-aware…but that it's not worth worrying about until we reach this point.

So what will happen as machines become more and more intelligent? At first, we can expect them to outperform us in any task where they have a clear advantage, like rote memory or mathematical calculation.

Some people predict that humans should use creativity to remain superior, but this is unlikely. Human intelligence is a combination of the intelligence of our bodies and brains—which has evolved over millions of years—and intelligence that we have decided to design into our brains, which is much more recent.

We don't yet know whether it will be easier for a machine or a human to make a new invention or devise a surprising artistic interpretation of some low-level data. But as both types of intelligence evolve over time, it is unlikely that they can remain complementary forever. The only way we can be sure to stay ahead of machines is by focusing on those abilities that they cannot match, such as our capacity for consciousness and introspection.

If we assume that machine consciousness will eventually resemble human consciousness, we can start considering what this might entail. It's worth bearing in mind that machine consciousness doesn't necessarily have to work in exactly the same way as ours; it just has to produce similar results when applied within any given domain such as scientific research or stock market trading.

The level of adaptability shown by our own brains when faced by new problems is one reason why many experts believe that current machines still lack a certain something even if they vastly outperform us in terms of raw data processing speed. For example, some AIs can now defeat the best humans at board games such as chess and Go, but this doesn't mean that they will be able to quickly adapt their strategies when playing a new opponent.

Another reason is that it's hard to set out clearly what consciousness actually consists of, let alone how it might work in an artificial context. So far, there are only loose prototypes for machine consciousness based on abstract ideas about consciousness rather than knowing what it is to be conscious in practice. But one thing seems certain: once we have created AI systems with capacities similar to our own, we may need to change our own consciousness in order to make the best use of it.

This raises the question: what kinds of artificial intelligence will need consciousness? After all, computers don't currently have it, and yet they are performing tasks that in the past only humans could do.

There are two reasons why people think that in the long term they will need consciousness to outstrip us at all tasks. The first is that they might have a capacity for introspection, which means that if their thinking gets stuck when designing something new or trying to solve a puzzle, then it might be helpful for them to think about what exactly they are thinking about. The second is that human consciousness comes with one big overwhelming advantage: the ability to empathise with other people. This means that we can understand what others are thinking, which helps us work together on tasks, negotiate agreements, and resolve conflicts. A conscious machine might be able to make itself more useful by developing this kind of access to the thoughts of other machines.

If this happens, it might be possible for us humans to accommodate our own consciousness alongside that of intelligent computers, perhaps by developing a third type of consciousness that is neither entirely human nor entirely artificial. This will be difficult but not impossible: for example, we already have abilities that do not come naturally to us, such as programming our own computer systems.

An important question that remains is whether it will always be necessary for intelligent machines to have consciousness or whether they might find other ways to collaborate with us. This would depend on how well they were able to model the behaviour of humans in different scenarios, and these simulations might already exist within restricted virtual reality worlds where computers can interact with each other without having access to any real data about the outside world. If this turns out to be possible, then it suggests there is no limit to human-computer synergy because artificial agents could work alongside people even if they did not need introspection or empathy.

That would be good news for us humans, because the alternative of living alongside self-conscious machines is enough to make anyone feel anxious.

Even though there are big debates about whether consciousness needs to be built into intelligent computers in order for them to outperform us, no one denies that it will give them some advantages over time. We may eventually need theories of machine consciousness in order to fully understand what kinds of intelligence they can develop and how we might work with them as a society. The only thing we know for sure is that someone's got to come up with those theories, or else we will not be able to compete with their superior thinking power. Perhaps the most important lesson from all this is that whatever works best, human consciousness or artificial consciousness will eventually become the only kind of consciousness that counts.

Artificial intelligence is emerging at an ever-increasing pace, and the technology of AI is constantly advancing. This shows no signs of slowing down, but with all advancements in new technologies come challenges for human beings to learn enough about them to make proper use of them.

This challenge presents itself as a need for researchers and developers to create ethical limitations for intelligent machines; we must ensure that they behave kindly toward humans and other forms of life and commit themselves fully to following our laws without exception (we don't want Skynet). There also exists the possibility for these powerful machines to replace us if we're not careful, which would be very dangerous indeed.

Although I'm sure you understand what this entails already, it's important to state that we will need new ways of explaining the decisions made by intelligent machines; we cannot leave these decisions up to human intuition, or we risk continuing to make them in an irrational and overly emotional way.

Instead, we must create hard constraints that describe the limits of intelligences' capabilities (and ensure that they behave ethically). We're asking for a lot, but we must balance the needs of people who demand ever-faster computers with those of people who need these computers to be ethical. I'm sure you agree this is necessary to ensure our continued survival as a species.

In conclusion: scientists have been studying artificial intelligence for decades now, and there've been many different strides forward in what it can do. It performs faster every day, and this continuous research ensures that it will continue to do so. This comes with a price, however: humans will need to ensure that these intelligent machines are capable of making rational decisions for us. We have the chance to build a better world with artificial intelligence, but we could also be giving our future robot overlords instructions on how to run it.

We need a solution that describes 'the limits of intelligences' capabilities' and ensures they behave ethically. We should not just leave the decisions up to human intuition. We should understand how we ourselves think first before passing those skills unto machines. They may be able to help us develop new technologies that have mind-boggling capabilities, but only if we can find a way of explaining what they're doing to people who don't have the technical skills needed.

We now face the challenge of maintaining our humanity as intelligent machines become more powerful with each passing day—I think you'll agree this is an important problem to solve.

ACKNOWLEDGEMENTS

Special thanks to:

My ever supportive and caring wife, Stella, for helping me birth this book, and being there for me through the ups and downs of the COVID-19 pandemic.

Vikas Malkani, my business and life coach, for your guidance and contribution of the 3Q Mental Model. May it be useful to everyone looking to harness the power of technology in a mindful and conscious way.

My co-writer, Pearlin Siow, and her team, who have spent so much time revising and editing this book. Your patience has been invaluable in incorporating the new ideas that came to me during our work on it!

My communications coach Sam Cawthorn, a superhuman survivor of a life-changing road accident who's overcome his disabilities. He's become an outstanding global keynote speaker and CEO of the Speaker Institute and been an incredible inspiration on this journey.

Leman Ng, my role model growing up. Thank you for infusing in me the habit of reading and helping me hone the art of building human relationships!

Thanks also to my three personal heroes, who inspire me every day to be a better leader and a wiser user of technology to empower my employees and customers. Industry leaders and giants all, they've transformed their respective industries through innovation, passion, and clear foresight:

- Anthony Tan, CEO of Grab. His company has not only changed the way we travel but made it so much more convenient. His work has truly added value for millions of people around the world.
- Tan Min-liang, CEO of Razer. He's known as the 'Steve Jobs of gaming', an epithet he's rightly earned for creating one of the most important gaming companies in the world from scratch. His perseverance and passion have created award-winning products used by leagues of gamers and video creators all over the world.
- Piyush Gupta, CEO of DBS Bank. It's Southeast Asia's largest bank in total assets and won Euromoney's World's Best Bank award in 2021. Under his leadership, it's grown from strength to strength, harnessing AI and digital technology to combine leading-edge banking services with a superior digital customer experience.

ABOUT THE AUTHOR

Tony Tan is the co-founder and CEO of Imperium, an AI and metaverse company that's received over twenty industry and business awards, including the prestigious Enterprise 50 Singapore. Imperium is a key partner of leading AI organisations including DataRobot, Splunk, and TigerGraph.

As an author and a business leader, he focuses on the future of digital innovation and disruptions. He is also a keynote speaker, a Speakers Institute facilitator, and a Speakers Tribe Singapore leader. In addition, he hosts the Tony Tan Show and Podcast.

Tony has helped over 300,000 corporate users from Fortune 500 companies to government agencies optimise their life and business results by leveraging the power of AI. He is the creator of EPIC.ai, which fuses the potential of AI and the creativity of human thought to create a digital 'influencer' that combines the strengths of both—creating powerful communication and connection, and optimised productivity.

ABOUT THE CO-AUTHOR

Boss of Me is a boutique book-writing agency run by Pearlin Siow that specialises in helping people write as well as publish books. Together with her team of content specialists, it has produced several bestselling biographies for top entrepreneurs and companies in Singapore. Their clients range from billionaires to stay-at-home mothers.

Connect with Pearlin at www.bossofme.sg.

www.ingramcontent.com/pod-product-compliance
Lightning Source LLC
Chambersburg PA
CBHW030944180526
45163CB00002B/696